Fort Larned

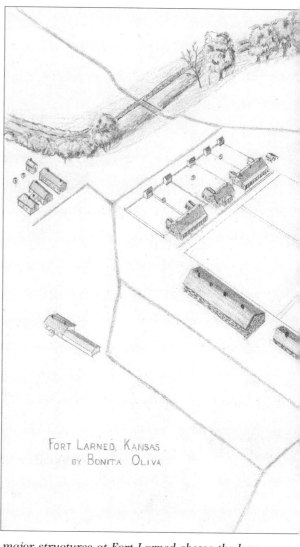

FORT LARNED, KANSAS
BY BONITA OLIVA

This aerial conception of the major structures at Fort Larned shows the location of the post in relation to Pawnee Fork, in back, and the dry oxbow, on right. This view is from the southeast and depicts the parade ground with officers' quarters on the west, enlisted men's barracks on the north, bakery and shops building and the new commissary storehouse on the east, and the old commissary storehouse and the quartermaster storehouse on the south. The blockhouse is below the southeast corner of the parade ground with a sentry box between it and the corner. One post trader's complex is shown southwest of the officers' quarters, and the other post trader's store is south of officers' row. The post

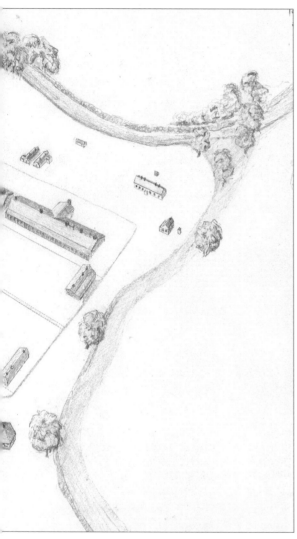

adjutant's office and office of the post commander is the small structure to the north of officers' row. Privies are behind each officers' quarters and enlisted men's barracks. The two sets of laundresses' quarters are depicted north of the middle of the barracks. The old adobe hospital and the hospital steward's quarters are shown north of the east end of the east barracks within which the post hospital was located after 1871. A bridge across Pawnee Fork can be seen behind the north officers' quarters. Several minor structures, stables, corrals, ice houses, and cemeteries have been omitted from this drawing. Sketch by Bonita M. Oliva, 1982.

Both Fort Larned and the city of Larned, Kansas, were named to honor Colonel Benjamin F. Larned, paymaster general of the U.S. Army when the post was established in 1859. Born in Pittsfield, Massachusetts, September 6, 1794, he entered the army during the War of 1812 as an ensign in the Twenty-first Infantry. He participated in the defense of Fort Erie and received the brevet rank of captain for gallant conduct. In 1815 he was appointed regimental paymaster. When the army was reduced and his regiment abolished, he was appointed paymaster of the Fifth Infantry with the rank of major. In 1847 Larned was appointed deputy paymaster general of the army with rank of lieutenant colonel. In 1854 he was promoted to the rank of colonel and appointed the paymaster general. He was eligible for retirement in the same year Fort Larned was founded, but he continued to serve. When the Civil War broke out he thoroughly reorganized his department to meet the needs of the enlarged army. He remained at his post until physically exhausted. He died in Washington, D.C., on the sixty-eighth anniversary of his birth, September 6, 1862. He was never in Kansas.

Fort Larned
Guardian of the Santa Fe Trail

by Leo E. Oliva

Kansas State Historical Society
Topeka, Kansas

THE AUTHOR: Dr. Leo E. Oliva is a former university professor of history. He farms with his wife, Bonita, in Rooks County, Kansas, and is the owner of Western Book publishing company. In addition Oliva is a freelance historian whose writing and research has focused on the frontier army and Indians as well as local history. This book originally was published in 1982 as part of a series Oliva prepared on Kansas forts for the Kansas State Historical Society. Similar books on Fort Hays (1980) and Fort Scott (1984) also were part of that series; revised editions of both books were published in 1996. Oliva's other publications include *Soldiers on the Santa Fe Trail* (1967), *Ash Rock and the Stone Church: The History of a Kansas Rural Community* (1983), and *Fort Union and the Frontier Army in the Southwest* (1993).

FRONT COVER: *The Escort*, Fort Larned, Kansas, 1859–1878, by Jerry D. Thomas. Thomas, a nationally acclaimed artist, has made a career of creating wildlife and western art. His original works will appear on the covers of all eight volumes of the Kansas Forts Series. Thomas is a resident of Manhattan, Kansas.

Fort Larned: Guardian of the Santa Fe Trail is the third volume in the Kansas Forts Series published by the Kansas State Historical Society in cooperation with the Kansas Forts Network.

Additional works in the Kansas Forts Series
Fort Scott: Courage and Conflict on the Border
Fort Hays: Keeping Peace on the Plains

Original title: Fort Larned on the Santa Fe Trail
Copyright © 1982 Kansas State Historical Society
Additional printings 1985, 1987, 1988, 1990
Revised edition 1997

Library of Congress Card Catalog Number 82-80495
ISBN: 0-87726-024-9

Printed by Mennonite Press, Inc., Newton, Kansas

Contents

FOREWORD

Following the United States acquisition of Louisiana from France in 1803, which nearly doubled the land area of the young nation, federal military policy in the Trans-Mississippi West focused on establishing a line of fortifications in advance of white settlement. The strategy was to provide a protective zone between Indians and white invaders, to control Indians and prevent them (and whites) from committing depredations, to enforce the various federal trade restrictions (including the sale of alcohol to Indians) in Indian country, and to maintain the legal integrity of Indian country as defined by the Indian Trade and Intercourse Act of 1834. Among the several important forts established during this period were Fort Snelling at the confluence of the Mississippi and Minnesota Rivers, Fort Leavenworth in present-day Kansas, and Fort Gibson in present-day Oklahoma.

The opening of the Santa Fe Trail trade in 1821, the beginning of large-scale white emigration to the Oregon country in the early 1840s, the annexation of Texas in 1845, the settlement of the Oregon question in 1846, the acquisition of the American Southwest in 1848 as a result of the Mexican War, and increasing conflicts with Indians between the Missouri River and the Rocky Mountains prompted a radical revision of western military and Indian policy in the mid-1840s. Militarily the new policy focused on control of Indians, while Indian policy shifted from the one-reservation "Indian country" theory to one of more limited, concentrated reservations. In this, Fort Larned—established as Camp on Pawnee Fork in 1859 and then Camp Alert prior to its permanent designation in honor of Colonel Benjamin F. Larned in 1860—played an extremely important role.

Located on the Santa Fe Trail near the confluence of Pawnee Creek and the Arkansas River, Fort Larned provided protection for trail commerce and the United States Postal Service; during the 1860s it served as headquarters and principal annuity distribution point for the Upper Arkansas (Cheyennes and Arapahos), and Kiowa, Comanche, and Apache Indian Agencies. It also provided military protection for federal land surveys, railroad construction crews, and Indian treaty delegations, and by the late 1860s had emerged as a major federal commissary for supplying the increasing number of Indian agencies in Indian Territory south of Kansas. It generally is recognized that Fort Larned was the most important federal military installation in western Kansas and that, in the Indian wars of the south-central Plains prior to the Medicine Lodge treaties of 1867, it was exceeded in importance only by Forts Leavenworth and Riley, both of which are located east of the ninety-eighth meridian.

Author Leo Oliva's analysis of military affairs along the Santa Fe Trail (1967) and history of Fort Union (1995) have established him as one of the most distinguished military historians of the American West. *Fort Larned: Guardian of the Santa Fe Trail* is a carefully compiled study of documentary evidence that is notable for its integration of social and economic data into the mainstream of Fort Larned's military role in the white advance onto the south-central Plains.

William E. Unrau
Endowment Association Distinguished Research
Professor of History Emeritus
Wichita State University

1

The Santa Fe Trail

Fort Larned was one of several military posts established to protect the Santa Fe Trail, which ran from Missouri to New Mexico, following the Arkansas River for a portion of that distance. The Santa Fe Trail was the oldest overland road across the Great Plains, connecting the Anglo and Spanish American cultures as well as crossing through that of the Plains Indians, when it was opened as a commercial route in 1821–1822. Traders from the new state of Missouri, led by William Becknell, were the first to succeed in transporting commodities to Santa Fe after Mexican independence from Spain in 1821. Becknell was accorded the title of "Founder of the Santa Fe trade and father of the Santa Fe Trail," and other traders followed his profitable path to New Mexico. Becknell took the first wagons to Santa Fe in 1822, opening the Cimarron Cutoff, which was used almost exclusively until the Mountain Branch was opened in 1845, after which both routes carried much traffic. Beginning in 1824 traders banded together in caravans for the trip to the Southwest, thereby providing protection from Indians. The U.S. government surveyed the Cimarron route in 1825. Later the army provided some protection from Indians.

Many Indian tribes lived in the region of the trail, and others came into the region to hunt buffalo periodically. Major tribes having contact with the route along the Arkansas River were the Pawnees, Arapahos, Cheyennes, Kiowas, Comanches, and Plains Apaches. Other tribes that were affected by the trail and were sometimes encountered by travelers

1

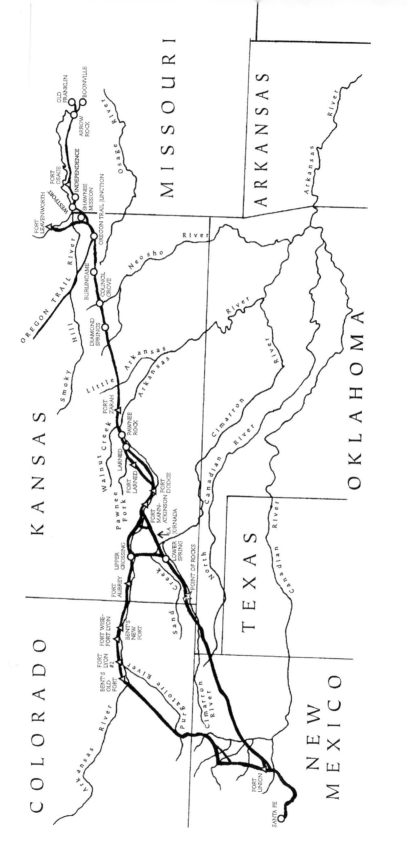

The Santa Fe Trail

included the Kansas, Osages, Jicarilla Apaches, and some bands of Sioux. All these tribes, even those who practiced some agriculture, depended on the buffalo for much of their food, clothing, and shelter. Horses were important to these Indians, and stealing horses was considered an honorable deed in their culture. Thus they were often a threat to travelers on the trail, seeking horses and other property. Travelers were simultaneously seen by Indians as a threat to their culture, especially after the 1840s when it was apparent that the buffalo on which their way of life depended was threatened by the invaders. Clashes were inevitable, and military protection of the Santa Fe route periodically was provided by the governments of Mexico and the United States during the commercial era.

The Santa Fe Trail was founded as a trade route, and it was essentially that until after the Mexican War, 1846–1848, when it became an immigrants' trail as well, leading to California and other gold fields as well as the new territory that became Arizona, New Mexico, Nevada, and parts of Utah and Colorado. During the early years the traders outfitted their wagon trains somewhere along the Missouri River, Independence and Westport becoming two major points of departure. From Independence to Santa Fe via the Cimarron Cutoff was approximately 770 miles, and the mountain route was about 825 miles.

Council Grove on the Neosho River became a major campsite, where caravans often were organized for the remainder of the trip. From that point the trail crossed several streams before joining the Arkansas River near present-day Great Bend. The trail followed the north bank of this river and crossed Walnut Creek, passed a famous landmark known as Pawnee Rock, and crossed Ash Creek before reaching Pawnee Fork of the Arkansas, also known as Pawnee River and Pawnee Creek.

An important campsite developed at the Pawnee crossing near which Fort Larned later was established. There the trail split into two branches, known as the Wet and Dry Routes. These routes rejoined at the point where Fort Dodge later was founded. From the Arkansas River there were several roads to the Cimarron River across approximately fifty miles of desert. Later the Mountain Branch followed the Arkansas into present-day Colorado, passed Bent's Fort (a famous trading post from 1833–1849 that was reconstructed by the National Park Service and is now a national historic site), and proceeded over Raton Pass into New Mexico to join the other route near where Fort Union (today a National Park Service-administered monument) was founded in 1851.

Trade caravans usually went to Santa Fe in the spring and returned to Missouri in the autumn, but there were exceptions. The major items

Wagons on the Santa Fe Trail often traveled four abreast, especially in Indian country, so a tight circle of wagons could be formed quickly in case of attack. The two outside columns would swing out and the lead wagons would move back toward each other, forming one half of a circle, while the two inner columns branched out to form the other half with the rear wagons closing the circle. The livestock could be kept within this protective circle, and the armed men could defend themselves while hiding behind the wagons.

taken to trade in New Mexico and beyond were cloth, hardware, cutlery, notions, and jewelry. Among the commodities brought back to the United States were gold, silver, donkeys, mules, furs, buffalo hides, wool, and Mexican blankets. The conquest of the Southwest by the United States in the Mexican War changed the trade from international to internal, but the trading patterns continued for some time. Trade increased as Anglo Americans moved to the region and wanted to maintain the same standard of living they had known in the East. The Anglos also wanted mail and stagecoach service, which was provided at least as early as 1849, and the railroad, which came in the 1870s. Increasing travel over the Santa Fe Trail and the establishment of stage stations was followed by increasing Indian resistance. Additional military protection was provided, and Fort Larned came into existence in 1859. By the end of the Civil War in 1865 forts had sprung up at several strategic points along the entire Santa

Fe road. A new mission was added for the troops when the Atchison, Topeka and Santa Fe Railroad line was constructed near the trail during the 1870s. With the arrival of the railroad in Santa Fe in 1880, the old trail temporarily fell into disuse. Most of the route became a part of the federal and state highway systems in the twentieth century. Long before that happened, however, the buffalo, Indians, and military posts were gone from the trail. Fort Larned was a part of that process, a federal government response to Indians.

2

The Founding of Fort Larned

Sporadic Indian raids on trader caravans characterized the early years of the Santa Fe Trail, but the post-Mexican War years saw increased resistance to expanded traffic. More military protection was provided to travelers, stage lines, and mail stations. Fort Atkinson was established in 1850 near the site of present-day Dodge City. This post was built of sod. Early in the first winter there, Captain William Hoffman, commanding the post, recommended that it be moved to Pawnee Fork where there was more timber and the location would be nearer Fort Leavenworth, the source of supplies.

In March 1851 orders were issued to move the post to Pawnee Fork, but this never occurred. A new commanding officer, Captain R.H. Chilton, was satisfied with the original site. Had that move been made, the site of later Fort Larned would have been a military post in 1851. Fort Atkinson was occupied until 1854 and then abandoned because of its isolated position and ineffectiveness in meeting Indian resistance. Even so, it had safeguarded a mail station for almost four years. The mail station was moved eastward to Walnut Creek, near present-day Great Bend, before the post was abandoned. It was the attempt to safeguard the mail route that led to Fort Larned's founding.

Although they had signed a peace treaty at Fort Atkinson in 1853, the Kiowas and Comanches became more serious threats to the trail in 1858 and 1859, probably because of the increased traffic of the Colorado gold

William W. Bent operated a large trading post on the Arkansas River in present-day Colorado. The first Bent's Fort, now a national historic site, was built in 1833 by William and his brother Charles. Bent's New Fort was opened in 1853. In 1859 William, who had a thorough knowledge of Plains Indians, was appointed Indian agent for the upper Arkansas River area. He recommended the establishment of a military post on Pawnee Fork just a short time before troops were sent there to protect the mail station and founded what became Fort Larned.

rush. By 1859 the value of goods shipped and equipment traveling over the trail was estimated at from three to ten million dollars annually. A Missouri newspaper reported that, between March 1 and July 31, 1859, 2,300 men, 1,970 wagons, 840 horses, 4,000 mules, 15,000 oxen, 73 carriages, and more than 1,900 tons of freight had moved westward on the trail. These estimates were considered conservative because the Colorado gold seekers were "too numerous to count." The Kiowas and Comanches simply took advantage of the opportunities available to them.

Their agent, Robert Miller, urged military punishment of the two tribes. The new agent, William Bent, appointed during 1859, recommended the establishment of a military post at Pawnee Fork and another near Bent's New Fort in Colorado as essential to protect the trail from the Kiowas and Comanches. His recommendations were put into effect within a year with the establishment of Forts Larned and Wise (later Lyon). In fact, before Bent made his recommendation, a series of events was under way that resulted in Fort Larned's occupation.

Because of the Indian raids, Captain W.D. DeSaussure and three companies of First Cavalry were sent from Fort Riley on June 10, 1859, to establish a summer camp near old Fort Atkinson and protect that sec-

Fort Riley, founded in 1853 at the juncture where the Smoky Hill and Republican Rivers form the Kansas River, was closely connected with Fort Larned. The troops that established Fort Larned came from Fort Riley, as did most of the supplies for the garrison at Pawnee Fork. During times of Indian troubles near Fort Larned, reinforcements usually came from Fort Riley. During the early years some troops from Fort Larned were sent to Fort Riley during the winter months because of better quarters and the reduced cost of supplying the men there. Fort Riley is still an active military post and has the U.S. Cavalry Museum. Drawing by John Gaddis.

tion of the trail. This force increased with the addition of another company of First Cavalry—under command of Captain W.T. Walker—during the summer. One of these companies was sent to Pawnee Fork to establish a camp that later became Fort Larned.

Pawnee Fork was considered as a site for a military post earlier without results. The camp located there in 1859 was occupied to protect a mail station. Mail contractor Jacob Hall attempted to build a station at Pawnee Fork in the spring of 1859, but the Kiowas and Comanches forced him to retreat with their threats to burn any buildings erected and to kill any men sent to build them at any point west of the mail station at Walnut Creek. Hall informed Postmaster General Joseph Holt of this problem and asked for government aid. Holt urged the War Department to provide necessary protection, and acting Secretary of War William R. Drinkard directed that troops be sent to Pawnee Fork.

Colonel Edwin V. Sumner, First Cavalry, issued the orders establishing Camp on Pawnee Fork in 1859. Sumner was an experienced officer with many years of service on the Plains. He had helped protect the Santa Fe Trail since the early 1840s.

Thus Captain DeSaussure was ordered to send one of his companies in the field near old Fort Atkinson to that place. Accordingly, on September 1, 1859, DeSaussure sent Captain Edward W.B. Newby and one company of First Cavalry to Pawnee Fork. With this assurance of military protection, Hall sent a wagon train of materials and sufficient employees under direction of William Butze on September 22 to build a mail station and corrals. Before this train reached Pawnee Fork, however, Captain Newby's company left camp there on September 17 in the company of the remainder of DeSaussure's command on the way to Fort Riley for the winter. Hall's repeated requests for protection combined with evidence of Indian hostility in the region resulted in troops being sent back to Pawnee Fork. One of the events that aided in this decision was a September 24 attack on the mail coach near Pawnee Fork by a party of Kiowas who killed two of the three-man crew.

Colonel Edwin V. Sumner, commanding the military department on October 4, 1859, ordered a company of First Cavalry to return and occupy a position near the mail station at Pawnee Fork. Thus seventy-five men of Company K, led by Captain George H. Steuart and Lieutenant

David Bell, arrived there on October 22. They encamped about one-half mile from the site of Captain Newby's summer camp, near the main trail crossing of Pawnee Fork and close to Lookout Hill (later known as Jenkin's Hill) approximately two and one-half miles downstream from the later, permanent site of Fort Larned.

The soldiers encountered no Indians on the trip out. One week after establishing Camp on Pawnee Fork, as it was officially known, some Kiowas camped about two miles downstream on the opposite bank. Two of the Indians were seen near the mail station the next day, and this information was taken to Captain Steuart. He sent Lieutenant Bell with a small detachment to investigate. They found two Kiowas and gave chase. The Indians showed signs of resistance, and both were killed. Steuart believed they were spies, trying to find out how many soldiers were at the camp. There was fear that the killings might touch off a general uprising by the Kiowas, but they did not. Steuart was able to devote attention to the protection of mail coaches passing in both directions.

In addition to protecting the mail station and escorting mail coaches along approximately 140 miles of the trail the small garrison was expected to build its own quarters and corral. These were constructed of sod, but some troops were quartered in tents and dugouts. By November 25 the corral was built, a supply of hay was cut and stacked for the winter, and some sod quarters were completed. Garrison supplies were shipped from Forts Leavenworth and Riley, and in November a wagon train brought four thousand rations and nine hundred bushels of corn.

The garrison was reduced in size in November because of inadequate quarters, the high cost of supplying the post, and the belief that Indians would be little trouble during the winter months. Captain Steuart and more than half the command went to Fort Riley, leaving Lieutenant Bell and thirty men at Pawnee Fork. Because of the shortage of forage, Captain Steuart took most of the cavalry mounts to Fort Riley and left Lieutenant Bell with five wagons and mule teams to pull them.

Lieutenant Bell immediately discovered that he had insufficient manpower to escort all the mail coaches, and he requested reinforcements. Lieutenant John D. O'Connell and twenty men of the Second Infantry were sent from Fort Riley with three wagons loaded with forage and rations. A civilian surgeon, A.L. Breysachre (also spelled Breysacher), accompanied them to the post where they arrived December 22, 1859. With these additional men Lieutenant Bell still was unable to provide an escort for all mail coaches.

11

These four sketches, above and on the facing page, drawn by Private Robert F. Roche, Company G, Second Infantry, who arrived at Fort Larned in May 1860, form a panoramic view of the post. This is the earliest known illustration of Fort Larned, drawn sometime between May and November 1860. The officers and men still lived in tents, but some of the adobe structures either were completed or under construction. Note the abundant timber on site at the time. Most of this was used to provide firewood to heat the quarters during the next few years, and later illustrations show few trees left. Private Roche also served as the hospital steward at Fort Larned from April 22, 1861, to October 31, 1862.

The mail station was safe. Indian resistance did decrease during the winter when they went into camps for the season. The garrison of approximately fifty survived the winter without major incident. Much effort was undoubtedly spent just keeping warm. On February 1, 1860, orders were issued changing the name of the post to Camp Alert because the soldiers had to remain constantly alert for Indians. This order reached the post on February 12.

Captain Henry W. Wessells, Second Infantry, arrived at Camp Alert on May 4, 1860, and assumed command of the post. He was accompanied by

13

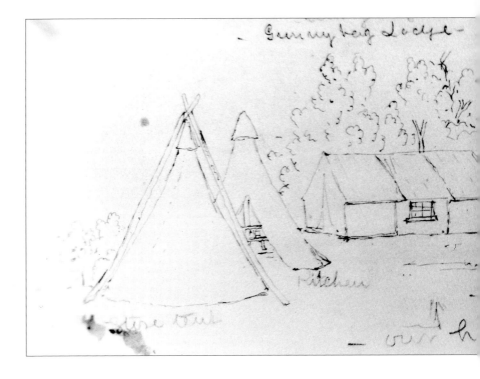

two companies of his regiment, which increased the number of men at the garrison to 160. He had orders to protect the mails and build a more permanent fort. Wessells selected a new site upstream from the camp about two and one-half miles, the present location of the post, and several adobe buildings were erected there during the summer of 1860. Wessells's request that the name be changed to honor the paymaster general, Colonel Benjamin Franklin Larned, became official on May 29, 1860.

The troops lived in tents while the adobe structures were being built. Civilian laborers helped with the construction. Most of these buildings were located outside the main quadrangle of sandstone structures, which were constructed between 1866 and 1868 and are still standing. The adobe buildings formed quite a complex when completed, but the material proved to be poorly adapted to the environment and deteriorated rapidly.

The adobe barracks and major storehouses were located north of the parade ground in two large structures, each approximately 210 by 24 feet. The west building housed one company of soldiers and the commissary storehouse, and the east one housed one company of troops and the

This sketch, also by Private Roche, shows in more detail his dwelling, kitchen tent, store tent, and an ambulance parked beside "our house."

quartermaster storehouse. Each barracks contained a kitchen and mess room. Two sets of laundresses' quarters were on the north side of the post—each seventy by eighteen feet with four rooms—where married soldiers lived and their wives served the post as laundresses. A hospital, also on the north side, approximately seventy-four by nineteen feet, contained a ward for sixteen beds as well as a kitchen and a dispensary. A bakery was located near the hospital, a dugout in the bank of the creek. The corrals were located east across a dry ravine, an old oxbow of Pawnee Fork.

The officers' quarters were housed in a 144-by-19-foot adobe building containing seven rooms and located south of the parade ground. Officers' stables were in dugouts in the creek bank. Shops for blacksmith, carpenter, and saddler were located in a fifty-by-twenty-foot building constructed of poles and canvas that stood southwest of the west barracks and commissary building. There was a twenty by twenty adobe storehouse, which temporarily may have been used as the post magazine. Another twenty by twenty adobe building served as the post guardhouse. Privies were located near the barracks, the hospital, and the officers' quarters. Other small buildings may have existed. At least two wells were dug to provide water for the post, but most drinking water was obtained from the creek because the underground water was strongly tainted with minerals.

Most of these structures were built hastily and poorly; adobe and sod (which was used in several of the roofs) were not permanent materials under such conditions. Repairs were made periodically, but the quarters never were very clean or comfortable. Whenever it rained muddy water dripped into the buildings. A much better complex of buildings was erected later when it was determined that Fort Larned would remain an active post after the Civil War. Nothing remains of these early buildings except for their foundations, most of which have been located.

While that construction was accomplished, the soldiers from Fort Larned continued to protect the mail station and escort stagecoaches. A military expedition sent from Fort Riley to deal with the Kiowas and Comanches during the summer of 1860 failed to accomplish that mission, but the presence of this force along the Santa Fe Trail helped to secure that route. The founding of Fort Wise (renamed Fort Lyon in 1862) in Colorado Territory near Bent's New Fort on September 1, 1860, also contributed to the security of the road on the eve of the Civil War. The Fort Larned garrison increased to 270 in September 1860 with the arrival of two companies of second dragoons. However, with the approach of winter in November the garrison was reduced to approximately sixty, and it remained at that level until the second year of the Civil War. The presence of Fort Larned and other military posts, as well as the activities of mail escorts and campaigns against Indians, combined to make the trail safe by 1861, when one officer reported that all was quiet in the Fort Larned region. The Civil War that erupted in the East in 1861 affected both the trail and Fort Larned.

3

Civil War Years

The Civil War affected Fort Larned indirectly for a time by forcing it to rely on volunteer replacements for the regular army troops who went to fight in the conflict. The war years saw increasing traffic over the trail, a further temptation for Indian raiders. As the Civil War dragged on and whites continued to fight each other, the Plains Indians seized the opportunity to increase their resistance, leading to more direct contact with the troops stationed at Fort Larned.

Captain Julius Hayden, Second Infantry, commanding officer at Fort Larned, reported that Arapahos were robbing wagon trains in the area during the summer of 1861. He had only insufficient numbers of infantrymen to deal with the mounted tribesmen. He provided escorts to coaches and wagon trains but could do little more without mounted troops, which were not available. This shortage of cavalrymen remained a problem. In May 1862 Captain Hayden reported that Kiowas, Plains Apaches, and Arapahos were attacking travelers east of the fort, and he requested mounted troops be sent to repulse them. The following month two companies of Second Kansas Cavalry arrived at the post, increasing the garrison from 63 to 292 and making it possible to send more patrols out to meet Indians. Additional reinforcements from the Ninth Kansas Cavalry followed.

The trail had protection, but Indians continued to stay near the fort and harass travelers at every opportunity. One year when Indians had gathered to await the distribution of their annuity goods their agent, S.G. Colley, came to Fort Larned and persuaded them to hunt away from the

17

Colonel Jesse H. Leavenworth, Colorado volunteers, commanded Fort Larned for a time during the Civil War in 1863. He worked hard to prevent Indians from closing the Santa Fe route during that conflict. After the Civil War Leavenworth became the government agent for the Kiowas and Comanches, with his offices at Fort Larned. With the agent for the Cheyennes and Arapahos, E.W. Wynkoop, he attempted to bring about a peaceful settlement of Indian problems. He participated in the peace treaties of 1865 and 1867 and resigned as agent during the winter campaign of 1868–1869.

trail until October 1, when they could return for their annuities. The Indians agreed and troubles ceased. In the autumn the Kansas volunteers returned to Fort Riley, which had better quarters and where the cost of supplying them was less. Indian troubles near Fort Larned ceased for the time being.

As usual, few Indian raids took place along the trail during the winter of 1862–1863. Indians began to reassemble along the route during April and May 1863, probably in anticipation of receiving their annuities. Their presence led to conflicts, and rumors spread that a major Indian uprising was imminent. Anticipating this, on June 8, 1863, Colonel Jesse

H. Leavenworth, Second Colorado Infantry, commander at Fort Larned, was placed in command of all troops serving along the Santa Fe Trail in the District of Kansas so he could coordinate military actions.

With so many more Indians than soldiers in the region Colonel Leavenworth feared any major uprising could not be repulsed. His fears were compounded on July 9 when a sentinel killed a Cheyenne at Fort Larned. The Cheyennes accepted the explanation that the man was intoxicated and failed to halt when ordered by the sentry as well as the gifts that were offered, and the feared retaliation never occurred. Colonel Leavenworth and the Fort Larned soldiers sighed with relief when the Indians left the area to hunt and raid along other sections of the trail. Indians returned to Fort Larned for annuities in the autumn and then went to their winter camps south of the Arkansas River. No serious Indian threat occurred through 1863, but that changed the following year when the long-awaited uprising became a reality.

By the late spring of 1864 Plains tribes were attacking settlements and travel routes in a determined campaign to drive whites from their country. First reports of attacks near Fort Larned came on April 20 when the Kiowas robbed some wagon trains. Lieutenant W.D. Crocker, Third Wisconsin Artillery, commanding at Fort Larned, declared he had not seen Indians so hostile in his two years at the post. He requested cavalry reinforcements, but mounted troops were not readily available.

Other tribes joined in the attacks, forcing more and more travelers to seek Fort Larned's protection. On May 17 the new commanding officer, Captain J.W. Parmetar, Twelfth Kansas Infantry, declared that "unless there is a cavalry force sent here, travel across the plains will have to be entirely suspended." This brought some cavalry from Fort Riley, and the mails and wagon trains resumed operations. Even then many employees at the stage stations and trading ranches were forced to abandon their places. Captain Parmetar took eighty men from Fort Larned and attempted to drive the Kiowas from the trail, but he failed to locate them. He later called a council of Indian leaders to meet near the fort and ordered them to leave the trail and go south for the summer. The Indians at the council declared that they were peaceful and agreed to go south, but while they were talking other members of their tribes were robbing trains west of Fort Larned. To assist the fort, ten companies of Colorado volunteers were stationed at Fort Lyon and Camp Wynkoop to protect the trail west of Fort Larned, and other troops were stationed at Council Grove and Salina to help east of the fort; without their help the trail probably would have been closed. Still Indian resistance continued.

On July 17, 1864, Kiowas raided Fort Larned and took 172 army animals. They were pursued but not overtaken. Raids against wagon trains continued in the region. Major General Samuel R. Curtis, departmental commander headquartered at Fort Leavenworth, concluded the Indian threat was so serious that he led a battalion of volunteer troops from Fort Riley to attempt to push Indians away from the Santa Fe Trail. This force, comprising four hundred men and two pieces of artillery, arrived at the Walnut Creek crossing on July 28. There General Curtis established Fort Zarah. He reported that Indians scattered as his force moved along the trail. He proceeded to Fort Larned where he split his command into three detachments and sent these north, west, and south of the post to punish any Indians found. No Indians were encountered by these forces because Indians would retreat rather than fight against strong odds. Thus the trail was temporarily safe, and Curtis returned eastward. Indians immediately resumed raiding wagon trains without troop protection. Whenever troops were sent in pursuit of Indian raiders the Indians seemed to disappear into the landscape. This led one military leader to describe them as being everywhere and yet nowhere, an extremely frustrating situation for the army.

Traffic on the Santa Fe Trail faced serious problems, due as much to fear of Indian attack as to actual Indian raids, as Colonel J.C. McFerran reported on August 28 after traveling from Kansas City to Santa Fe:

> Both life and property on this route is almost at the mercy of the Indians. Every tribe that frequents the plains is engaged in daily depredations on trains, and immense losses to the Government and individuals have occurred, and many lives have already been lost. Several persons were killed and large numbers of animals run off during my trip. . . . Many contractors and private trains are now corralled and unable to move from their camps for fear of Indians, and other trains have had their entire stock run off, and cannot move until other animals can be had.
>
> This evil is on the increase, and the number of troops on the route is so small that they are unable to securely protect the public property at their respective stations. They have in several instances lost a large number of public horses and other animals, run off by these Indians, within a few hundred yards of their posts. Soldiers and citizens have been killed within sight of a large number of troops. You cannot imagine a worse state of things than exists now on this route. Women and children have been taken prisoners to suffer treatment worse than death.

In September General James Blunt, commander of the military district, took charge of a body of troops at Fort Larned and marched west-

Major General Samuel R. Curtis led a military expedition from Fort Riley in the summer of 1864 to prevent Indians from closing the Santa Fe Trail. He established Fort Zarah and visited Fort Larned. The Indians fled as his force marched along the trail only to return after Curtis returned to his headquarters in Fort Leavenworth.

ward, looking for Indians. About seventy-five miles west of the fort on September 25, Blunt attacked an encampment estimated to contain more than three thousand Kiowas, Arapahos, and Cheyennes. He routed and pursued them for several days, reportedly killing nine and wounding many more. The Indians finally escaped, and the troops—two of whom were killed and seven of whom were wounded—returned to Fort Larned. Thwarted by that expedition and the approaching cold weather, Indian raids in Kansas decreased, but hostilities in Colorado increased. There a punitive expedition under the command of Colonel John M. Chivington, Third Colorado Cavalry, attacked a Cheyenne and Arapaho encampment on Sand Creek, November 29, 1864. Those Indians claimed to be at peace, but their village was destroyed in a fierce attack that is still the subject of controversy. Reports of Indian losses there varied from approximately seventy-five to more than five hundred. After this attack the Santa Fe Trail became quiet, and only a few incidents took place during the winter months.

On January 17, 1865, approximately forty-five Cheyennes and Arapahos attacked a sutler's train west of Fort Larned, but the escort forced

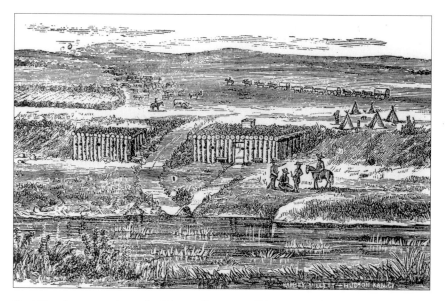

Fort Zarah, approximately thirty miles east of Fort Larned, was established by General Samuel R. Curtis on July 28, 1864, and named for his son, Major Zarah Curtis, who was killed at Baxter Springs. Located approximately two miles from the Arkansas River near the Santa Fe Trail crossing of Walnut Creek, the first quarters were dugouts and tents. Until June 30, 1868, it was not classified as a separate fort but was under the commanding officer of Fort Larned and made its reports to that officer.

the Indians to retreat after three were killed and three were wounded. On February 1 fifteen Indians charged a fatigue party of eight soldiers chopping wood near Fort Zarah and killed one man. But such incidents were few during the winter months.

Military leaders, fearing renewed Indian resistance along the trail in the spring and summer of 1865, set up a new system of escorting wagon trains, established three new military posts along the trail—Camp Nichols and Forts Dodge and Aubrey—and prepared to conduct a major campaign against any hostile tribes. At the same time new efforts to negotiate a peaceful settlement with the Plains tribes were undertaken. The result was greater safety for travelers on the trail in 1865.

Fort Larned was selected as the east end of an escort system with Fort Union, New Mexico, the western point. Beginning March 1 an escorted train departed from each point on the first and fifteenth day of every

An Indian agency for the Cheyennes and Arapahos was located at Fort Zarah in 1865–1866. During 1865 a little blockhouse, guardhouse, and sutler's store were added, as was a bridge across Walnut Creek. These structures, along with a trader's ranch, appear in this sketch by A. Hunnius, June 27, 1867. The major purpose of the garrison, which fluctuated between 52 and 641, was to provide escorts. During a typical month, January 1865, the troops from Fort Zarah provided twenty-one escorts who traveled a combined distance of twelve hundred miles although never going farther than forty-five miles from the post.

month. All merchants were expected to wait for the military escort. East of Fort Larned wagon trains were held up until a large caravan with at least one hundred armed men could be organized to protect itself between Fort Larned and Council Grove. All government trains between Fort Larned and Council Grove were to have military escorts, and merchants were welcome to accompany them. The escort system was extended eastward from Fort Larned to Council Grove on May 1. No trains were permitted to leave between those dates, which helped reduce the number of Indian raids considerably.

The peace efforts were pursued earnestly during 1865, but a large military force was held in readiness at Fort Riley in case Indian hostilities increased. This force moved to Forts Zarah and Larned several times during the summer when it appeared that peace efforts were failing, but each time the troops were withdrawn.

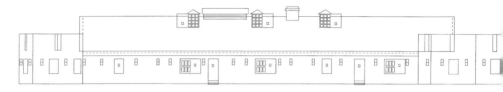

Fort Zarah was moved to a point closer to the Arkansas River in December 1867, where a large stone structure was completed to serve the needs of the garrison. Captain Almon F. Rockwell, quartermaster department, oversaw the construction of this building at the same time he was directing the erection of new stone buildings at Fort Larned. This sketch made from a blueprint prepared by First Lieutenant M.R. Brown, Corps of Engineers, chief engineer, Department of the Missouri, in 1867, shows the new Fort Zarah building that housed officers and troops and contained kitchen, mess hall, and storage facilities. The post was abandoned by order of October 15, 1869. Nothing remains at either site of Fort Zarah.

In August many leaders of the Plains tribes met with civil and military leaders at the mouth of the Little Arkansas River (present-day Wichita) and agreed to stop all raiding and remain at peace. They were to meet again in October to sign a treaty settlement. These treaties of the Little Arkansas did not bring lasting peace to the Plains, but they were a hopeful sign. The Civil War was over and the Indian war temporarily was halted. Fort Larned and the trail entered into a new, post-war era with promises of peace that slowly soured because of the tremendous influx of travelers and settlers into Indian country and renewed efforts on the part of the Plains tribes to preserve their hunting culture. Fort Larned was rebuilt during that time.

4

Building Fort Larned

The sod and adobe structures at Fort Larned were replaced by durable sandstone buildings after the Civil War. What was a set of rather miserable and uncomfortable quarters became a handsome and comfortable post, one of the finest on the Plains. It is only because of these stone buildings that anything of Fort Larned exists today. The first stone building was not a government structure but a post sutler's store, built in 1863 by the firm of Weischselbaum and Crane. Located west of the adobe officers' quarters, this store contained a saloon. Sometime later an addition was built to house a billiards room.

The first stone military structure was a blockhouse. Following the Kiowa raid of July 17, 1864, when 172 horses were captured from the post, an army inspector recommended that a blockhouse be built to provide protection from Indian attack. On July 31 General Curtis issued an order requiring all military posts to provide protection, such as a stockade, for all livestock and troops, and singled out Fort Larned for not having a stone blockhouse or a stockade for the animals.

A hexagonal blockhouse was completed on February 20, 1865. This building, twenty-two feet on each side, had one hundred portholes from which soldiers could shoot while protected. The building included a small cellar from which an underground passageway ran to a well nearby. The original roof, supported by the side walls and a large post in the center, was made of poles, brush, hay, and sod. This was replaced in 1867 by a wood-shingled roof topped by a watchtower.

General James G. Blunt led an expedition from Fort Larned in September 1864 that attacked a large Indian encampment west of the post. This was one of the few such expeditions during the Civil War that captured a camp of Indians before they were able to scatter in all directions and escape without detection.

Fort Larned was not attacked by Indians and the blockhouse probably never was used for its intended purpose. It served several other functions. Quartermaster supplies and ammunition were stored there for a time, at least until the new quartermaster storehouse was built in 1867. After that the blockhouse was utilized as a guardhouse because the old jail was too dilapidated to hold prisoners. The portholes on those walls that had prison cells were plastered shut in 1870 after it was discovered that liquor was being passed through them to the prisoners.

A new complex of nine sandstone buildings was constructed between 1866 and 1868. These were built on four sides of a four-hundred-foot-square parade ground. All remain today and are being restored as nearly as possible to their original condition. At the center of the parade ground was a wooden flagpole that was more than one hundred feet high. It was brought overland in sections from Fort Leavenworth and erected similarly to a ship's mast. This flagpole was struck by lightning and destroyed on May 12, 1877. The large garrison flags flown had a thirty-six-foot fly and twenty-foot hoist. From this pole the flag was visible for many miles as one approached Fort Larned.

Colonel John M. Chivington, Third Colorado Cavalry, was a Methodist minister who joined the Union forces in Colorado when the Civil War commenced in 1861. He served with distinction in New Mexico when Confederate troops were turned back near Santa Fe. He achieved undying infamy in 1864 after leading an attack on a peaceful village of Cheyennes and Arapahos located on Sand Creek in Colorado. The controversy over that massacre continues to the present day. A congressional investigation found Chivington's actions dishonorable, but he escaped punishment by resigning from the military. When the Indians later retaliated against the whites, the troops as well as the Indian agents at Fort Larned were in the thick of military action and peace negotiations.

The new buildings were constructed of sandstone quarried nearby. At least three known quarry sites are within three to six miles of the post; one of these, about three miles downstream and near the original site of Camp on Pawnee Fork, is still visible. The stone was prepared and laid up by professional stone masons, civilian employees hired by the army. Civilian carpenters, plasterers, and laborers also were hired to provide most of the labor of construction, although soldiers were detailed to such work when possible. Wages were high. Stone masons and plasterers

Colonel Christopher "Kit" Carson, New Mexico volunteers, visited Fort Larned following the signing of the treaties of the Little Arkansas, at which he served as one of the peace commissioners. Regarding his visit, a reporter for The Plains newspaper at the fort declared: "There is no more faithful or valuable officer to the government on the frontier than Kit Carson."

received ninety dollars per month; carpenters, eighty-five dollars; and laborers, thirty-five dollars. A private's pay at the time started at sixteen dollars per month. Of course the soldier had all food, clothing, and shelter supplied, but the army also provided board and room for the civilian employees. They lived in tents and dugouts. The quartermaster officer who oversaw much of this construction was Captain Almon F. Rockwell.

The first of these stone structures was a commissary storehouse located at the east end of the south side of the parade ground. Completed in 1866, it probably was constructed first because the adobe storehouse reportedly had insufficient capacity and also was in such a state of disrepair that it no longer protected commissary supplies. Because it was on the most exposed side of the post, portholes were formed in the south wall from which soldiers could fire their weapons if the post were attacked. Several photographs show that a wooden lean-to was constructed on the east end of this warehouse; its purpose is unknown.

Work began on two enlisted men's barracks during 1867. The same year work began on three officers' quarters, a quartermaster storehouse and a shops building (including a bakery). These were all completed early in 1868. Since Captain Rockwell directed most of this construction program, and

Captain Almon F. Rockwell, quartermaster department, was sent to Fort Larned during 1867–1868 to oversee the construction of many of the stone buildings at the post. He also directed construction of a new stone structure at Fort Zarah in 1867. He was the only officer from the quartermaster department ever stationed at Fort Larned. The fact that all buildings constructed at Fort Larned under his direction still stand attests to his skills.

many of the craftsmen worked throughout the period, it is understandable that many workmanship similarities appear in the buildings.

The two enlisted men's barracks on the north side of the parade ground were each 161 by 43 feet, with kitchens later attached to the north side of the buildings. Each contained quarters for two companies—one company located at each end—with orderly rooms, mess rooms, and kitchens for both companies. Storage cellars were under the kitchens. Each company quarters, designed to house approximately eighty men, was contained within a forty-foot-square room with ten-foot-high ceilings. Double two-tiered bunks slept two up and two down. There were windows for ventilation, and the rooms were heated by wood-burning stoves. If companies were full these quarters were crowded, but most of the time companies were not at full strength. The garrison's size was so reduced by the early 1870s that the post hospital was moved into the east end of the east barracks, which was remodeled for that purpose in 1871. In 1875 Fort Larned soldiers received foot lockers for storing their wearing apparel and personal items. Prior to 1875 each soldier's knapsack was packed with his effects and kept on the shelf at the end of his bunk.

This sketch of Fort Larned in 1867 by A. Hunnius shows the fort during the era of new construction. It appears that the adobe barracks, left, and officers' quarters, right of center, were still in use. A number of troops were housed in tents. The blockhouse is located in back just right of the flagpole; the first stone commissary storehouse is beside it on the right. Note the absence of trees.

This sketch of Fort Larned by Theodore R. Davis appeared in Harper's Weekly, *June 8, 1867, and illustrates that some men still lived in dugouts along the banks of Pawnee Fork.*

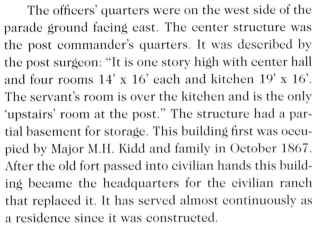

The officers' quarters were on the west side of the parade ground facing east. The center structure was the post commander's quarters. It was described by the post surgeon: "It is one story high with center hall and four rooms 14' x 16' each and kitchen 19' x 16'. The servant's room is over the kitchen and is the only 'upstairs' room at the post." The structure had a partial basement for storage. This building first was occupied by Major M.H. Kidd and family in October 1867. After the old fort passed into civilian hands this building became the headquarters for the civilian ranch that replaced it. It has served almost continuously as a residence since it was constructed.

The other officers' quarters, one on each side of the commanding officer's quarters, were identical structures. Each was a single-story rectangle with kitchen wings attached in rear at the ends. Captains' quarters had two rooms, a servant's room, and a kitchen, while lieutenants' quarters consisted of a single room, although two lieutenants usually shared two rooms. In 1870 frame additions were added to the lieutenants' quarters giving them a shared kitchen, dining room, and servant's room. All officers' quarters were heated by wood-burning stoves.

In addition to the post commander's quarters, Fort Larned had quarters for four captains and eight lieutenants. During the peak years of occupation these were sometimes inadequate, forcing some officers to crowd together or live in tents. At all military posts officer housing was selected by rank, and any officer who outranked another could "bump" the lower-ranking officer from his quarters. With the high turnover among officers at the post, considerable moving occurred. Each officer who was bumped could bump any lower-ranking officer; the results resembled musical chairs. If all officers' quarters were occupied, the lowest-ranking officer might be found living in a tent.

The quartermaster's storehouse was erected on the south side of the parade ground west of the com-

These two sets of enlisted men's barracks stood on the north side of the parade ground. Each had kitchens attached in the center rear and each contained quarters for two companies of troops, although the east portion of the east building was converted to a hospital in 1871. Both have been restored to this appearance.

A photograph of the rear of the barracks and the kitchens attached. Storage cellars were located under the kitchens.

The commanding officer's quarters, located in the center of the west side of the parade ground, was a roomy and comfortable dwelling. The servants' quarters above the kitchen can be seen at left rear. The sketch below illustrates the original floor plan of the building.

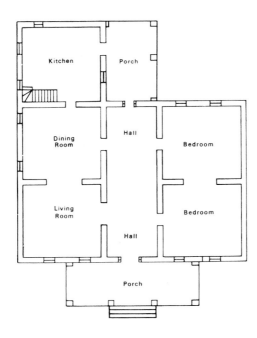

Major Meredith Helm Kidd, Tenth Cavalry, commanded Fort Larned from May 1867 to March 1868. He was the first commanding officer to live in the new stone commanding officer's quarters, which were completed in October 1867. At the time Major Kidd was thirty-six years of age and his family included his wife, Millicent, and three children: Rose, age eight; Edmund, age five; and Lelia, age two. He had been a lawyer in Indiana and had spent five years in the California gold fields during the 1850s. He became a soldier in an Indiana regiment during the Civil War. He rose to the position of chief of artillery on Brigadier General Jeremiah Sullivan's staff. In 1864 he was appointed a major in the Eleventh Indiana Cavalry and saw action in Tennessee and Mississippi before being sent to Kansas where he served on the Santa Fe Trail in 1865. He was promoted to lieutenant colonel prior to his discharge in 1865. He then returned to law practice in Indiana and published a newspaper. In 1867 he was offered the rank of major in the Tenth Cavalry, a black regiment with white officers, and was sent to Fort Larned. After leaving Fort Larned, Major Kidd served at other forts in Kansas and Indian Territory, including Fort Wallace and Camp Supply. In 1870 he was charged by a superior officer, Colonel Benjamin H. Grierson, as being unfit for duty and an inefficient commander. Kidd resigned from the service even though he was found competent to command by an army board. He returned to Indiana for a short time, traveled to Texas where he bought a herd of cattle, and drove the cattle to Kansas. In 1873 he went back to Indiana and practiced law. He later worked with Indian tribes in Oklahoma Territory and Colorado. He died in 1908 in Indiana.

missary storehouse. It measured 158 by 40 feet and contained an office for the post quartermaster, an issue room, a bedroom for the quartermaster clerk, and a large storage room. This large room often was too small for all the materials handled by the quartermaster's department, which included all military clothing, camp and garrison equipage, wagons, draft animals, forage, fuel, and materials for construction and repair. This building had iron bars over the windows to safeguard the property.

In addition to being the army's storekeeper the quartermaster was responsible for the repair shops at the post. He hired civilian workers for his department, and was responsible for the upkeep of all military buildings.

The shops building was a one-story rectangle, eighty-four by thirty feet, built on the east side of the parade ground at the north end. It housed a bakery, a saddler shop, a wheelwright and carpentry shop, and a blacksmith shop, which had at least one forge. The bakehouse had one oven with a capacity of 340 rations per day. The bakery was a responsibility of the commissary officer, and bakers were detailed from enlisted men of the garrison. Baking knowledge was not a qualification for the job.

Officers' row, viewed here from the south in 1879, stood on the west side of the parade ground. Note the single rail fence and boardwalk that extended along the front of the three buildings. The large porches added a touch of elegance to the otherwise plain but substantial dwellings. Privacy fences screened off the backyards from the rest of the post. The backyards contained privies, water barrels filled with water hauled from the creek, and stables for the officers' horses.

This sketch represents the floor plan, not to scale, of the two junior officers' quarters: (A) servant's room, (B) kitchen, (C) captains' quarters, (D) lieutenants' quarters, one to each room, and (E) hall. The plan does not include additions made in 1870.

Front view of one of the junior officers' quarters. These quarters were used as dwellings during the ranching period.

In addition to supplying the bread needs of the post, the bakery was a source of income for company and post funds. A company was permitted to sell unused portions of its bread issue, and the money went into a company fund that was used to purchase items not supplied by the army, such as fresh vegetables or fruit, eggs, a water cooler, or newspapers and books. Anyone at the post not entitled to draw a bread ration was per-

This 1867 photograph shows all the stone buildings erected at the post except the new commissary storehouse, which was completed in late 1868. This view is from the east, with the commanding officer's quarters across the parade ground directly behind the flagpole. The other officers' quarters do not have the full-length porches, which are shown in all later photographs. The sutler's store is partially hidden behind the large tree trunk left of center. Note the sparsity of trees.

This photograph of Fort Larned was taken from the top of the blockhouse in the late 1870s, probably 1877 or 1878. The sentry box is at lower left. The old commissary storehouse and quartermaster storehouse are on the left, and officers' row is across the parade ground. The end of the new commissary storehouse can be seen at the right.

37

Another view from the top of the blockhouse shows the remainder of the new commissary warehouse on the left, shops and bakery building on the right, and the enlisted men's barracks at the center.

Fort Larned had several wells, although the drinking water was obtained from Pawnee Fork because the groundwater was sulphurous. This well stood behind one of the barracks. The building to the left was the old adobe hospital, used in the 1870s as the ordnance sergeant's quarters and magazine. The house to the right of the well was the hospital steward's quarters.

mitted to buy bread, the money going into the post fund for use in buying items such as band instruments, musical scores, and books.

The last stone structure built at Fort Larned was the new commissary storehouse erected in 1868. After it was built the first stone commissary building was designated the "old commissary storehouse." Captain Rockwell left Fort Larned before this building was completed. It was used for purposes other than foodstuff storage. The ordnance officer used part of it as a magazine at times, for the fort never had a separate stone facility for ammunition. This new commissary also was used as an emergency hospital and isolation ward from time to time prior to 1871, when the east end of the east barracks was converted into the post hospital. The post chaplain may have held Sunday services in the new commissary, and a post school for soldiers' children was ordered by the commanding officer in October 1871 to be held in the north end of the building. The post chaplain, officers' wives, and others took turns teaching school.

Other buildings at Fort Larned included stables, ice houses, quarters for the hospital steward, and an adjutant's office at the northwest corner of the parade ground. After 1868 the post had the appearance of a small town, and it was a comfortable as well as a functional complex. Newspaper reporter Henry M. Stanley visited Fort Larned in April and October 1867. After his second stop he observed: "A complete change had been effected at Fort Larned. . . . The shabby, vermin-breeding adobe and wooden houses have been torn down, and new and stately buildings of hewn sandstone stand in their stead." He also noted that the "comfort of the troops has been taken into consideration by the architect and builder." When compared with some other frontier outposts, Fort Larned was indeed an imposing place.

A military reservation four miles square was declared and surveyed in 1867, but civilian settlers were not always kept out of this reservation as required by law. After the post was abandoned the reservation area was surveyed again, this time to comply with public land delineations, and settlers were given title through the General Land Office.

5

Frontier Defense

Fort Larned was established and existed for almost two decades because of Indian resistance to travel and mail service over the Santa Fe Trail. Indian relations occupied much of the post garrison's time during its early years. No major battle occurred near the post, and the fort was never attacked by Indians. The troops there did help protect the trail and the Atchison, Topeka and Santa Fe Railroad—which followed the route—and they assisted with the removal of Plains tribes from the Arkansas valley in Kansas. The post also served as a distribution point for tribal annuities and was the home of Indian agents for a time. This resulted in an unusual twist of circumstances; the fort that was established to keep Indians from the trail became a gathering place where large numbers of Indians came to collect their annuities.

The protective missions provided by the fort from its founding through the Civil War continued after 1865. Military escorts for travelers, protection of mail stations, campaigns against Indians, and peace negotiations continued. Fort Larned subsequently became the distribution point for annuity goods to Cheyenne and Arapaho tribes from 1862 until 1868. This practice almost resulted in conflict in August 1862 when some Cheyennes and Arapahos threatened to seize their annuities stored at Fort Larned and scheduled to be delivered to them in October. Traders apparently urged the tribes to attempt the seizure so the Indians would have more goods with which to trade. The plan was thwarted by the army, and the Indians were persuaded to go hunting until October. The

Cheyenne Indians were one of the major groups opposing travelers on the Santa Fe Trail, the construction of the railroad through their hunting grounds, and the military sent to stop their raiding. Depicted here in front of their dwellings are Cheyenne tribal members Eagle Shirt and Black Horse.

following year a large band of Kiowas came to the post, proclaimed they wanted to trade, and then drove off three hundred beef cattle.

In August 1863 Little Heart, a Cheyenne, believed to have been intoxicated and on his way to the fort to purchase more whiskey, failed to halt when so ordered by post sentry Isaac Marrs. Marrs shot and killed Little Heart. There was great fear at Fort Larned that the incident might incite the large number of Indians encamped nearby to initiate a major conflict. Fortunately, the Cheyennes were placated. Later in 1863 the agency for Kiowas, Plains Apaches, and Comanches was established at Fort Larned.

As a by-product of the Civil War, a group of Caddo Indians settled near Fort Larned during 1863. Approximately one thousand Caddoes, who were farming near Fort Cobb in present-day Oklahoma, were abandoned by their agent who joined the Confederacy. The Caddoes supported the Union and came north seeking aid and a place of security. The

Bureau of Indian Affairs sent five thousand dollars to agent S.G. Colley to relocate and help sustain these people. The money was used to help them begin farming along the south side of Pawnee Fork, where two thousand acres were surveyed for their use. There they built homes and planted approximately 250 acres of irrigated corn in the spring of 1864; the Caddoes left the area during the fall of the year, fearing they would be caught up in the Indian war on the Plains. What they left behind soon was appropriated by Indians and whites. This was believed to be the first experiment with irrigated farming in the valley, a practice that became widespread in the twentieth century.

Following the Indian war of 1864 and the treaties of the Little Arkansas in 1865, Indian resistance along the Santa Fe route significantly was reduced during 1866. The 1865 conference expanded Fort Larned's agency role. While Jesse Leavenworth served as an agent for the Kiowas and Comanches at Fort Larned another agency under E.W. Wynkoop serving Cheyennes, Arapahos, and Plains Apaches was established first at Fort Zarah and later at Fort Larned. These agencies operated until 1868 when the Indians were located on reservations in present-day Oklahoma and the agencies were relocated at Fort Cobb.

Not all tribesmen had agreed to the terms of the 1865 treaties, and others were determined not to abide by their terms. Thus Indian hunting and raiding along the Santa Fe route renewed, especially by a group of Cheyennes known as dog soldiers. Although it was relatively peaceful during much of 1866 near Fort Larned, Indian raids in Texas and on the Northern Plains gave rise to fears that the Indian war soon would be renewed in Kansas. Rumors that an Indian uprising was planned for the spring of 1867 caused military leaders to withhold arms and ammunition from Indians, causing the Indians to conclude that the treaty promises made them were not being honored by the United States. The arms and ammunition were promised in return for the peace settlement.

In February 1867 Indian interpreter Fred Jones claimed that Kiowa chief Satanta had told him that the Indians wanted all military posts removed from the Plains, Santa Fe Trail traffic to be stopped at Council Grove, and the railroad to be stopped at Junction City. If this were not done, the Indians would combine and drive the whites from the region. Such reports—even if untrue as this one turned out to be—convinced military leaders that an Indian uprising was imminent. In response a military campaign was undertaken in the spring of 1867 to attempt to defeat those Indians who had not taken up residence on their assigned reservations.

Black Kettle
Cheyenne

Little Raven
Arapaho

Satank
Kiowa

Satanta
Kiowa

Among the Indian leaders of the tribes in the Fort Larned vicinity were these men, above and on the facing page. All were at Fort Larned at one time or another. Annuities were distributed to their people from the fort during the 1860s. Some of these chiefs were involved in raids during which livestock was stolen at Fort Larned. Several were at the post during the preliminaries of the Medicine Lodge peace council in 1867. All were valuable leaders, and military officials and Indian agents respected them.

Ten Bears
Comanche

Yellow Bear
Arapaho

Major General Winfield S. Hancock, commanding the military department, organized a force of fourteen hundred troops, including four companies of the newly organized Seventh Cavalry under Lieutenant Colonel (Brevet Major General) George Armstrong Custer at Fort Riley, to march along the Santa Fe Trail and deal with Indians as necessary to enforce the treaties. Among the civilian scouts were James Butler "Wild Bill" Hickok, Jack Harvey, and Tom Atkins.

Hancock led his force through Fort Harker and Fort Zarah to arrive at Fort Larned on April 7, 1867, where he hoped to meet with Indian leaders. Agents Leavenworth and Wynkoop made arrangements for a council, but a snowstorm hit on April 9, one day before the planned meeting, and several days passed before anyone could move. When the Indian leaders did not come to the fort, Hancock marched his command up Pawnee Fork to where some Cheyennes and Sioux were encamped. The approach of the large body of soldiers frightened the women and children, and the encampment hastily was abandoned. Hancock was convinced that the Indians must have been hostile or they would not have fled. He sent Custer and the Seventh Cavalry in pursuit. Custer failed to overtake the Indians, but he reached the Smoky Hill Trail to the north and found that Indians had attacked stage stations there. Assuming that theses attacks were carried out by Indians who had fled Pawnee Fork camp, Hancock ordered the village, tipis, and all other property burned. His orders were carried out on April 19. Custer pursued the Indians until

Edward W. Wynkoop (1836–1891), left, posing with "Texas Jack" Crawford, was Indian agent for the Cheyennes, Arapahos, and Plains Apaches, 1865–1868, first as special agent with offices at Fort Zarah and later as regular agent with offices at Fort Larned. Wynkoop was one of the founders of Denver, and he served in the First Colorado Cavalry during the Civil War. As commander of Fort Lyon in 1864, he accepted the peaceful declarations of a group of Cheyennes and Arapahos at face value, for which he was removed from command. The Indians concerned later were attacked by Chivington at Sand Creek. Wynkoop was outraged by the Sand Creek massacre, and he later opposed General Hancock's destruction of the Indian village on Pawnee Fork. He participated in the treaty negotiations at the Little Arkansas River in 1865 and Medicine Lodge Creek in 1867. He was unable to restrain some of his charges in 1868, who went on a raiding tour through portions of Kansas. He resigned as agent in protest to General Sheridan's winter campaign. The agency was moved from Fort Larned to Indian Territory at the same time. Wynkoop was a consistent defender of Indian rights during his years as agent. He was particularly important in the pacification of Plains tribes during 1865–1866.

General Winfield Scott Hancock, who compiled a brilliant Civil War record, was department commander over the region including Fort Larned in 1867. He led an expedition against the Indians in the spring of that year, which brought him to Fort Larned where he met with Indian leaders. His expedition was not successful because he did not understand Indians. His actions, especially burning an Indian village on Pawnee Fork, increased Indian hostilities in the area. He was replaced later in the year by General Philip H. Sheridan. Hancock was the Democratic candidate for president in 1880 and was defeated by James A. Garfield.

July, when he abandoned his command at Fort Wallace in western Kansas and made an unauthorized trip to Fort Riley to see his wife. His actions led to his arrest, court-martial, and conviction; he was suspended from duty and pay for one year.

Hancock, after destroying the village, marched the remainder of his command to Fort Dodge, where he met with several Kiowa chiefs, including Satanta. Hancock was so impressed with Satanta's promises of peace and good will that he presented the chief with a major general's uniform, which Satanta reportedly was wearing later when he drove off the livestock at Fort Dodge. Hancock gave up his campaign and returned to his Fort Leavenworth headquarters. His destruction of the Indian village probably contributed to the increase in Indian resistance during the summer of 1867. Soldiers at Fort Larned and other posts were kept busy with escort duties and investigating reports of Indian attacks.

The renewed warfare prompted Congress to create in July 1867 the Indian Peace Commission, which was charged with bringing peace to the Plains. This commission concluded that war was inevitable as long as Indians occupied lands that whites wanted. The solution was

Lieutenant Colonel (Brevet Major General) George A. Custer, Seventh Cavalry, spent one week at Fort Larned as part of Hancock's expedition in April 1867. After Cheyennes and Sioux abandoned their camp on Pawnee Fork, Custer was sent in pursuit. He spent much of the summer searching for but not finding the Indians. He abandoned his command in July and went to Fort Riley to see his wife. This led to his arrest and subsequent court-martial. He was found guilty on several counts and suspended from command without pay for one year. He returned to duty before that year was over because General Sheridan wanted Custer to lead his regiment in the winter campaign of 1868–1869. Custer contributed to the defeat of the Southern Plains tribes with his victory at the Washita in present-day Oklahoma, November 27, 1868.

to settle Indians on reservations away from white settlements and routes of travel. The results were the Medicine Lodge Creek council and treaties of October 1867. Preliminary talks were held between Indian leaders and the Peace Commission at Fort Larned, where they agreed to meet at Medicine Lodge Creek. A wagon train loaded with gifts for the Indians and supplies for the conferees was sent from Fort Larned to the treaty site. The treaties, signed with chiefs of the Cheyennes, Plains Apaches, Kiowas, Comanches, and Arapahos, provided that all warfare would stop immediately, all offenders against whites and Indians would be punished, and reservations would be provided for the tribes in present-day Oklahoma. The Indians agreed to withdraw all opposition to military posts, wagon roads, railroads, and promised not to hunt north of the Arkansas River. When this treaty was enforced the military need for Fort Larned would cease. But all Indians were not yet ready to concede the end of their way of life, and white buffalo hunters slaughter of the Indians' major

Theodore R. Davis was an artist correspondent for Harper's Weekly and Harper's New Monthly Magazine during and after the Civil War. He accompanied the Hancock expedition in 1867 and spent several days at Fort Larned. His reports and illustrations of the army and the Plains Indians were major sources of information for eastern readers.

Henry M. Stanley, an eastern newspaper correspondent, accompanied General Hancock during the expedition of 1867, which brought him to Fort Larned in April. He visited the post again in October 1867 prior to the Medicine Lodge peace council, which he attended. Stanley's reports gave eastern readers a firsthand account of the army and Indians. He was much impressed with the new construction at Fort Larned in 1867, but he described another frontier fort in Kansas as appearing like a "giant wart on the plains." In 1869 the New York Herald sent Stanley to Africa to find David Livingstone. He achieved lasting fame when he found Livingstone in 1871. Sketch from Harper's Weekly, July 27, 1872.

This sketch by Theodore Davis depicts Hancock's burning of the Indian village on Pawnee Fork, April 19, 1867. Sketch from Harper's Weekly, *June 8, 1867.*

resource brought additional Indian retaliation. By 1868 the Kansas Pacific Railway, fifty miles north of Fort Larned, took most of the traffic off the Santa Fe Trail near the fort.

In the spring of 1868 many Indians, for whom reservations were not yet established, came to Fort Larned to draw supplies from their agents. These supplies had not yet arrived at the post, so the hungry Indians began begging food from travelers and hunting north of the Arkansas River. Some of the more dissatisfied Indians began attacking white settlements in Kansas, and Major General Philip H. Sheridan, who had replaced Hancock as department commander, determined that the Indians would be defeated and forced to stay on their reservations. Leavenworth and Wynkoop resigned as agents in protest of Sheridan's declaration of war against their charges. On September 25, 1868, Fort Larned ceased to be an agency site and distribution center for Indian annuities. Fort Cobb, Indian Territory, was designated for those purposes since it was close to the reservations.

Sheridan planned a winter campaign against the Indians and made preparations to march three columns of troops into Indian Territory. Several battles were fought, including Custer's defeat of Black Kettle's Cheyenne village at the Washita, November 27, 1868, and the Indians were forced onto the reservations. There would be later conflicts, but

Council at Medicine Lodge Creek, October 1867. Sketch by J. Howland in Harper's Weekly, *November 16, 1867.*

none would involve Fort Larned. Custer's suspension was terminated at Sheridan's request so he could participate in the campaign.

Fort Larned was occupied for another decade, but there was little to do during that time. By 1871 no escorts were furnished to wagon trains or other travelers, although railroad survey and construction crews were given military protection. In December 1871 a band of Kaw Indians who had been hunting buffalo without success came to Fort Larned and requested rations so they could return to their reservation near Council Grove without starving.

During the active period of Indian conflicts at Fort Larned, 1859–1869, according to newspaper files, approximately two hundred Indians and whites died in conflict in the region served by the fort. About the same number were reported wounded, bringing the total casualty count to almost four hundred. If those figures are correct, and they are difficult to verify, the losses averaged fewer than four each month for more than a decade. The property loss would have been much greater because theft rather than killing was often the Indian raiders' objective.

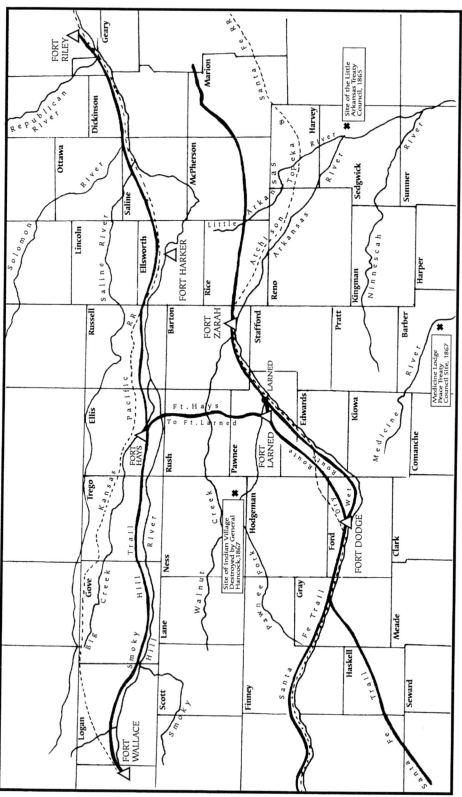

The Fort Larned section

When one considers the thousands who traveled the Santa Fe Trail and the millions of dollars worth of merchandise transported over the route, the losses seem small when compared with modern warfare.

Fort Larned's presence on the trail and the distribution of annuities to Indians there undoubtedly made a significant contribution to safe passage of most people and goods. Fort Larned apparently did more to keep the peace than to make war. The Indians of the Plains were not defeated by the military, although the army contributed to their decline. The Indian way of life was destroyed by the railroads, settlers, and loss of the buffalo. The technology and sheer numbers of white settlers rendered the buffalo-horse culture of the Plains obsolete and gave Indians little if anything to take its place. Fort Larned was a part of that transition; therein lies the significance of this small post on the Santa Fe Trail.

6

Officers and Troops

During its two decades of occupation, Fort Larned was commanded by forty officers, several of whom served more than once. During the Civil War most officers were not regular army but were members of state volunteer regiments. No famous officer ever commanded the post; the two best known were Majors John E. Yard and Richard I. Dodge. (For a complete list of commanding officers, *see* Appendix.)

The garrison usually comprised both cavalry and infantry companies sometimes supplemented with an artillery battery. Cavalry was not at the post after 1869. The number of troops stationed there varied considerably. During the early years the summer garrison often reached 150 officers and men, but during the winter months it was reduced, sometimes to as few as thirty, because of inadequate winter quarters and the high cost of supplying the garrison. The troops would be more comfortable and could be maintained at less expense if they were removed to Fort Riley during that time. The garrison also fluctuated during the Civil War because of troop shortages on the national level and because of reinforcements being sent to Fort Larned during times of Indian troubles. Thus the garrison might have 100 to 150 troops for several months then jump to 400 or 500 for a short period of time and fall back to 150 again. In the spring of 1865 six hundred were stationed there. During the years of Indian troubles after the Civil War, the garrison strength ranged from 150 to 300 much of the time, but still dropped lower at times. After 1868 manpower gradually declined, and by the last year of

Captain Henry W. Wessells
Second Infantry
1860

Captain Scott J. Anthony
First Colorado Cavalry
1864

Captain William H. Backus
First Colorado Cavalry
1864

Captain Thomas M. Moses
Second Colorado Cavalry
1865

Captain Theodore Conkey
Third Wisconsin Cavalry
1865

Pictured here are some of the commanding officers at Fort Larned. For a complete list of commanding officers, see Appendix.

Major Hiram Dryer
Thirteenth Infantry
1865–1866

Major Cuvier Grover
Third Infantry
1866

Major John E. Yard
Tenth Cavalry
1868–1869

Captain Simon Snyder
Fifth Infantry
1872, 1873–1874

occupation the garrison had only a portion of one company of troops, Nineteenth Infantry, which consisted of approximately thirty-five officers and men. After Fort Larned was deactivated in 1878 a small detach-

An 1867 photograph of Company C, Third Infantry, in front of their barracks at Fort Larned. Lieutenant Lorenzo Wesley Cooke, front center, was company commander. This is one of the few known photographs taken of troops at the post.

ment of troops was left to guard government property until the military reservation and post buildings were transferred to the General Land Office in 1882. It must be remembered that Fort Larned was a base of operations, not a true fortification, and troops were shuttled to and from as well as through the post as they were needed at various trouble points on the Plains.

Men from eleven regular army regiments and thirteen volunteer regiments were stationed at Fort Larned. The units with the longest service at the post were Third Infantry (almost six years), Nineteenth Infantry (more than four years), Second Colorado Cavalry (almost three years), Second and Fifth Infantry (more than two years), Tenth Cavalry (twenty-one months), Ninth Kansas Cavalry (more than sixteen months), and Second U.S. Infantry (thirteen months). The other sixteen units were present less than one year. The constant turnover of officers and men at the post made it unlikely that many post traditions were established. All this reflected the fact that Fort Larned was a small post destined for a temporary existence.

Thomas H. Allsup, shown here in later life, was a private in Company A, Tenth Cavalry, stationed at Fort Larned, 1867–1869. The Tenth Cavalry, known as "buffalo soldiers," comprised black enlisted men commanded by white officers. Allsup served at many western forts before retiring from the service as a first sergeant in 1897. His grandson, Thomas Allsup III, participated in ceremonies at Fort Larned National Historic Site in 1980, commemorating the service of black troopers on the frontier.

The troops from the Second U.S. Infantry, stationed at Fort Larned in 1865 and 1866, were former Confederate soldiers who were captured and held as Union prisoners. They were permitted to leave prison if they volunteered for service in the West for the Union. Known as "galvanized yankees," they were stationed at posts such as Fort Larned to help protect men and property from Indian raids. They served well in this capacity.

Black soldiers served at Fort Larned from 1867 to 1869, when men of the Tenth Cavalry were part of the garrison. Known as "buffalo soldiers," these black troopers and their white officers compiled an excellent record. Black units had the lowest desertion rates and the least incidence of alcoholism in the army. Black troops often faced discrimination and hostility at military posts in the West, and those who served at Fort

Private Flavius J. Glenn
Company D, Fifteenth
Kansas Cavalry
1865

First Sergeant LeRoy N. Buell
Forty-eighth Wisconsin
Infantry
1865

Private Ado Hunnius
Company D, Third Infantry
1867–1868

Private Charles Abbott Kennedy
Company K, Third Infantry
1867–1870

Commissary Sergeant
James Ryan
1873–1878

A few of the enlisted men who served at Fort Larned.

Richard Irving Dodge graduated from West Point in 1848 and remained in the service until retirement in 1891. He was a major in the Third Infantry when he commanded Fort Larned in 1871. He served at a number of western forts and was a student of the region and Indians. He published The Plains of the Great West *(1877) and* Thirty-three Years Among Our Wild Indians *(1882).*

Larned were the objects of discriminatory practices by officers and enlisted men of other regiments. Since each company was housed and messed separately, little direct contact took place between blacks and whites in quarters. At work and leisure, however, there were contacts and a few conflicts. An incident at the sutler's store in December 1868 almost resulted in a race riot. The black regiment's cavalry stables mysteriously caught fire and burned during the early morning of January 2, 1869. Believed to have been arson, the losses included the stables, 39 horses, 30 tons of hay, 40 saddles, 500 bushels of grain, and 6,000 rounds of ammunition. To defuse the situation post commander Major Yard sent the sole black company to Fort Zarah. No further racial problems were reported.

Although no famous military leaders were stationed at Fort Larned, several visited there. J.E.B. Stuart was at the post in 1860 before he became a famous Confederate cavalry leader. At the time he was a lieutenant in the First Cavalry and part of Major John Sedgwick's expedition against the Kiowas. John M. Chivington and Samuel R. Curtis were at the fort during the Civil War. Winfield Scott Hancock and George A. Custer were there for a week in 1867. Inspector General Randolph B. Marcy inspected the post in 1867 during the era of construction. William S. Har-

First Lieutenant
Frank D. Baldwin
early 1870s

Second Lieutenant
Charles E. Campbell
1869–1872

Second Lieutenant
Samuel N. Crane
1864

First Lieutenant
Samuel T. Cushing
1861

Captain Charles Ewing
1865

Lieutenant
Albert C. Gooding
1865

Among the many junior officers who served at Fort Larned are the men pictured here, pages 62-64. Many of the photos were taken after they had served at this post, and therefore they are both older and hold higher ranks than when they were at Fort Larned. The ranks listed are those held while serving at Fort Larned.

Captain John McLean Hildt
1866–1869

First Lieutenant
Charles H. Hoyt
1865

Captain Ezra W. Kingsbury
1865

First Lieutenant
Stevens T. Norvell
1865–1866

First Lieutenant
George F. Raulston
1867–1868

Captain Silas S. Soule
Civil War

Second Lieutenant
George K. Spencer
1874–1875

Captain William Steele
1860

Captain
Thomas G. Townsend
1871–1872

General Philip H. Sheridan, famous Civil War commander, replaced Hancock as commander of the Department of the Missouri in late 1867. He organized the winter campaign against the Indians' camps in present-day Oklahoma, 1868–1869. He was at Fort Larned in 1868 to meet with Kiowa and Comanche leaders. Sheridan's plan to destroy the Indians' winter camps proved to be a major step toward ending Indian domination on the Plains. Sheridan became general-in-chief of the army in 1883 and held that post until his death in 1888.

ney, C.C. Augur, Alfred Terry, and John B. Sanborn were at the post as part of the Indian Peace Commission that negotiated the Medicine Lodge treaties in 1867. William B. Hazen came to Fort Larned in 1868 to arrange for carrying out the provisions of the Medicine Lodge treaties. Philip H. Sheridan was there in September 1868 to council with Kiowas and Comanches and try to persuade them to stay out of the Indian conflict that year. John Pope inspected Fort Larned in July 1870. He condemned the adobe hospital but would not authorize repairs or a new hospital until he knew whether Fort Larned would be kept an active post. In the meantime patients used the new commissary storehouse as a temporary hospital ward when new recruits arrived in August and came down with fevers.

It was not famous officers who built the post and provided services to travelers, mail coaches, and railroad crews. That work was performed by enlisted men, many of Irish and German ancestry, whose names are known only in the records. It was these men serving the nation at a frontier outpost in an isolated region with inadequate accommodations and pay who deserve recognition for the daily duties they performed and the successful completion of the missions assigned to Fort Larned.

7

Economic Concerns

From an economic standpoint, Fort Larned's neighbors benefited from its presence. The soldiers' pay—which at Fort Larned amounted to thousands of dollars each month—was only a small portion of the money that benefited the local economy. How enlisted men and officers spent their pay is difficult to accurately determine, but that most of it ended up in merchants' hands (including the post sutler, general stores in nearby communities, saloons, and bordellos) is certain. The pay for an enlisted man after 1871 ranged from thirteen dollars per month for a new private to about thirty dollars for a noncommissioned officer with several years of service. Officers' pay began at a base of $150 per month during the era Fort Larned was occupied, and some officers received expense accounts in addition to salaries.

Since most of the soldier's necessities were furnished by his employer, his pay was available for "luxuries." If a soldier hired a company laundress to keep his clothing and bedding clean, the fixed charge ranged from one to two dollars per month for all laundry except an overcoat and bedsack, for which there was an additional charge of twenty-five cents each. Laundresses charges could be avoided if the soldier took his own hands and clothes to the washboard. Some soldiers cut each other's hair, saving the twenty-five cents charged at a barbershop. What monthly pay remained could be used for a variety of purchases; rarely does it appear that any was saved, although many soldiers sent some of their pay home to their families.

Civilian scouts were hired to assist the army and were at or near Fort Larned at various times. This sketch from Harper's Weekly, *June 29, 1867, is of George A. Custer's scouts, William A. Comstock, Edmund Guerrier, Thomas Atkins, and Thomas Kincade.*

The troops usually were paid in cash every other month, but on occasion, such as during the Civil War, the paymaster's visit came only once in four or even six months. This made payday an even more important event to celebrate, and the accumulated pay of several months gave the soldiers a fairly large sum with which to slake the thirst of a long dry spell. The spree after payday often brought the work of the garrison almost to a standstill. This caused debate among military leaders regarding pay frequency.

In 1867 the adjutant general reported: "It has been suggested by many intelligent officers that more frequent payments would tend to diminish brawls and desertions and temptation to intemperance, by keeping the men more constantly supplied with such small sums as they need for the moderate wants of themselves or families, instead of throwing in their hands comparatively large sums, sometimes the accumulated pay of six months." The inspector general argued the other side: "The

Civilian Scouts

Dick Parr

James B. Hickok

William F. Cody

The post trader's store was the only civilian establishment permitted on a military reservation. The post trader or sutler (as he was known then) had a contract with the army to operate a store in return for a specified annual payment. Prices charged at the store were determined by a board of officers who permitted a certain markup over cost of the items sold. Fort Larned enjoyed both good sutlers and sutlers' stores. This stone store was built in 1863 by Crane and Weichselbaum. It later had a billiard room attached. A frame mess house and sutler's home also were added. In 1867 a frame building was erected by another sutler, E.S.W. Drought; it had a bowling alley attached. In addition to the stores themselves, the sutlers had a complement of other buildings. For example, in 1866 the sutlers' property included a residence, mess house, two stables, ice house, carriage house, chicken house, and smoke house. From 1867 to 1869 two sutlers' stores operated at the post. They had an officers' room where men could drink, play cards, shoot billiards, and visit. They also had a place where enlisted men could purchase beer and liquor, gamble, and relax. The store, open to everyone, offered a wide choice of items. The following is a partial list of items and prices at Fort Larned in 1863, illustrative of the kinds of products offered and the cost of living at a western army post during the Civil War years (the sizes of cans, bottles, and boxes listed are unknown):

Potatoes, per bushel	$2.25	Beer, per gallon	$1.00
Apples, per bushel	3.50	Whiskey, per gallon	1.50
Flour, per sack	4.75	Crackers, per pound	.13
Tomatoes, per can	.60	Corn Meal, per pound	.04
Peaches, per can	.85	Butter, per pound	.25

Strawberries, per can	.85	Chocolate, per pound	.60
Oysters, per can	.75	Brown Sugar, per pound	.18
Lobsters, per can	.50	Cheese, per pound	.22
Pineapple, per can	.85	Tea, per pound	1.25
Jelly, per can	.75	Chewing Tobacco, per pound	.90
Coffee, per box	.45	Mixed Candy, per pound	.60
Clothes Pins, per box	2.50	Soap, per pound	.30
Cigars, per box	4.50	Playing Cards, pack	.25
Eggs, per dozen	.30	Diaper Pins, @	.25
Tomato Catsup, per bottle	.20	Neckties, @	.30
Castor Oil, per bottle	.25	Candles, @	.25
Cologne, per bottle	.25	Wash Boards, @	1.00
Blue Jean Pants, per pair	4.75	Hoop Skirts, @	2.50
Canvas, per yard	.25	Lead Pencils, @	.10
Blankets, @	$11.00-15.00	Smoking Pipes, @	$1.10-7.50

Other items included song books, fishhooks, coffee pots, guitar strings, saddles, lanterns, Epsom salt, cloth, pots and pans, hats, matches, needles and thread, spices, nails, revolvers, buttons, sulphur, hair dye, turpentine, wallets, tin buckets, molasses, axes, padlocks, scissors, mirrors, beads, and horse liniment.

---

longer you keep soldiers' pay, the longer you prevent them from desertion, and the more seldom they are paid, the fewer drunken irregularities will occur." The paymaster general objected to frequent payments on the grounds that the soldiers would never have enough money at one time to send any home to their families. The pay schedule remained unchanged, and the soldiers continued to spend their pay at local establishments whenever they had it.

In 1870 Fort Larned surgeon James N. Laing treated a fatal case of diarrhea that was attributed to drinking heavily of "the stuff sold as whiskey." Dr. Laing deplored the intoxication and demoralization that followed the appearance of the paymaster every two months. To offset the seriousness of inebriations following the sixty-day "dry spells," he advocated paying the troops every month. No changes were made.

Because of insufficient military manpower to provide all services needed by the army, such as scouting, hauling supplies, herding livestock, cutting firewood, repairing buildings and equipment, keeping records, and similar tasks, civilian employees always were present at Fort Larned and other military posts. At Fort Larned the number of civilians hired averaged fewer than ten per month except during the years of

Theodore Weichselbaum, right, and his partner, Jesse Crane, were appointed sutlers at Fort Larned in 1859 and opened for business there in 1860. They also had contracts as sutlers at other military posts. They purchased the goods and transported them to the post, where hired clerks ran the day-to-day business. Crane sold out his share of the Fort Larned store in 1866 to John E. Tappan who, with Weichselbaum, continued the operation until 1869.

Nels Cederberg was clerk in the Fort Larned sutler's store from 1866–1869.

George Crane, brother of sutler Jesse Crane, served as clerk at the Fort Larned store during the early 1860s.

Henry Booth served as post sutler at Fort Larned, 1869–1873, and was a founder of the city of Larned. Booth was born in England, May 11, 1838, and came to America with his parents at an early age. The family lived in Rhode Island, which Henry left in 1856 to settle on 160 acres near Manhattan, Kansas. He volunteered for military service in Kansas infantry and cavalry units during the Civil War. He rose to the rank of captain in the Eleventh Kansas Cavalry and participated in several battles in Arkansas. In 1864 his regiment returned to Kansas, and Booth commanded Fort Riley for a brief time before his assignment as chief of cavalry and

inspecting officer for the military district of the Upper Arkansas, including Fort Larned. While on an inspection tour, Captain Booth and another officer became separated from their escort near Fort Zarah and were attacked by twenty-eight Indians. They escaped, but Booth was twice wounded and his partner received four wounds. It was reported that twenty-two arrows were removed from the wagon in which they were riding. He saw duty in Wyoming Territory before his discharge in 1865. Booth returned to Manhattan and operated a hardware and farm implement business. In 1867 he was elected to the Kansas legislature, and in 1869 he became post trader and postmaster at Fort Larned. He assisted with the founding of the city of Larned and moved one of his frame sutler buildings from the fort to the town, reportedly floating it part of the way on Pawnee Fork. This building was credited with being more than just the first building in the town; it was the first saloon, restaurant, church, school, and hotel. He left his job as post trader in 1873 and built a home in Larned. He was elected to the state legislature from Pawnee County in 1873 and served as chief clerk of the Kansas House of Representatives during 1875–1876. In 1878 he was appointed receiver of the U.S. Land Office in Larned, in which position he assisted later with the sale of public lands of the military reservation after the fort was abandoned.

E.S.W. Drought served in the military during the Civil War and was appointed postmaster at Fort Larned in 1867. He was a contractor and builder, and he built a second sutler's store at Fort Larned in 1867. He served there as a trader and postmaster until December 1869 when he moved to Leavenworth. He later served as a sheriff of Wyandotte County, Kansas, and served in the state legislature. He built the Wyandotte County courthouse in the early 1880s and constructed many schools and businesses in that area.

major construction from 1866 to 1868, when the number of civilians hired reached more than two hundred and the monthly civilian payroll exceeded ten thousand dollars. This money was dispensed into the local economy through a host of clerks, guides, teamsters, laborers, herdsmen, watchmen, blacksmiths, wheelwrights, stone masons, carpenters, and others. The need for civilian workers declined after construction was completed, but a few remained until the post was deactivated.

In addition to salaries, large sums were spent to purchase army supplies in the region. Contracts with private firms that supplied hay, wood, beef, and other needs were awarded throughout the era of Fort Larned's occupation. Whenever possible, fresh vegetables, grain, milk, eggs, and other items were purchased from the surrounding community to save transportation costs. There were few local suppliers around Fort Larned until after 1870, but in the 1870s local farmers benefited from the fort as marketplace. All items that could not be found locally had to be shipped in at great expense. The troops at the post were urged to grow their own vegetables in company gardens, but these often ended in failure because of drought, neglect, insects, or a combination of these.

Military supplies usually reached Fort Leavenworth on the Missouri River by steamboat, and these were hauled from that point by wagon to Fort Riley and on to Fort Larned until the railroad built westward. Freighting military supplies was big business, for the freight charges often were greater than the cost of the products being freighted. The numerous employees of freighting firms such as Russell, Majors and Waddell held jobs because of the army, and many of those employees later became settlers in the West.

With the completion of the Union Pacific Railway, Eastern Division (later the Kansas Pacific Railway) to Hays in 1867, supplies were sent to that point by rail and hauled by wagon the sixty miles to Fort Larned. The Atchison, Topeka and Santa Fe Railroad, which arrived at the new town of Larned eight miles from the fort in 1872, derived much of its business from the army. It also was protected by the soldiers. The railroad was welcomed by the troops because it provided speedy transportation for men and supplies. In November 1872 soldiers from Fort Larned took wagons to the town of Larned to take delivery of four thousand pounds of fresh vegetables that had arrived on the new railroad.

CIVILIAN EMPLOYEES AT FORT LARNED, SEPTEMBER 1867
as reported in monthly post returns

No.	Occupation	Rate of Monthly Pay	Duties
1	Chief Clerk	$125	In charge of office
1	Clerk	$100	Commissary & QM offices
1	Clerk	$100	Personnel reports
1	Clerk	$100	Property & storekeeper
1	Master Mechanic	$100	Supt. of building const.
1	Interpreter	$100	With Indian agency
5	Blacksmiths	$85	Shoeing public animals, repairing wagons, and manufacturing iron work for buildings
4	Foremen	$60	In charge of various parties of laborers constructing new buildings
134	Laborers	$35	Quarrying stone, etc., for buildings, assisting masons, digging foundations, etc.

2	Wagon Masters	$75	Public transportation
2	Asst. Wagon Mast.	$45	" "
58	Teamsters	$35	" "
1	Ambulance Driver	$35	Conveying sick
30	Carpenters	$85	Construction and repairs
2	Saddlers	$90	Repairing harness
41	Stone Masons	$90	Construction
4	Plasterers	$90	Construction and repairs
3	Wheelwrights	$90	Repairing wagons
1	Painter	$75	Painting bldgs. & wagons
1	Tinner	$75	Repairing water spouts, etc.
1	Sail Maker	$75	Repairing wagon covers, tents, etc.

295 total employees $15,830.00 total salaries

Civilian post sutlers operated a general store on the post and provided many commodities and services to military and civilian personnel. The sutler's store contributed to the economy because the commodities had to be transported to the post and clerks had to be hired to run the store. From 1867 to 1869 two sutler's stores operated at Fort Larned. Usually one of the post traders served as the postmaster at the fort, another federal office that brought government money to the area. In addition to post sutlers were regimental sutlers who accompanied troops in the field and to posts where the regiment was stationed. There also were stagecoach offices connected with the sutler's store or independent of it at the post.

In these and other ways the federal money spent on a military installation bolstered the regional economy. Then, as now, it was economically desirable to have an army base nearby. Because of the economic benefits it is understandable that communities located near a post wanted the garrison to remain long after the troops had fulfilled their mission. This was true for Fort Larned, which was maintained for several years after the Indians had departed from the region.

8

Life at the Fort

The lives of Fort Larned soldiers were ones of isolation, although there were opportunities for contact with travelers on the trail, at whiskey ranches and houses of prostitution nearby, and after 1872 in the town of Larned. The soldiers' duties were monotonous and living conditions were not always desirable. Housed first in tents, sod huts, dugouts, and adobe buildings, the soldiers had neither comfortable quarters nor many conveniences. The construction of stone buildings in the late 1860s changed that for the better. The food was often of poor quality during the early years because it had been transported by wagon over vast distances, and it was sometimes spoiled from improper storage and the length of time in transit and storage before consumption. Sometimes the salt pork was rancid, and the flour often had worms in it.

The soldiers' diet had little variety. Hash, stew, bread, salt pork, and beans were standard items, and coffee and sugar were included in the rations. Fresh meat often was available, sometimes from freshly slaughtered beef and sometimes from buffalo. Most of this meat was served in stew, usually accompanied by some canned vegetables. Each company was responsible for feeding its own troopers, and soldiers ate at a company mess hall located behind each barracks. Meals were prepared by cooks of various talents since they were selected from within the company without regard to previous experience. While at the post, the men also had the daily bread from the post bakery. Troops in the field prepared their own mess individually and generally had poorer quality vittles except for the fresh game they hunted. The lack of fresh vegetables in the diet was a problem, and scurvy was a disease that plagued the troops in those early years.

Although there were few places for the soldier to escape from the routines of garrison life, there were saloons on and off post, and "hog ranches" that supplied prostitutes as well as whiskey just off the military reservation. After the railroad came the town of Larned was a few miles downstream. Boyd's Ranch—perhaps the most famous hog ranch—offered plenty of entertainment near the original site of Camp on Pawnee Fork. A pass was required to leave the post and visit any outside establishments, but sometimes soldiers were found absent without leave trying to escape the boredom of post life.

Soldiers were punished for breaches of military regulations, including absence without leave, sleeping while on duty, theft, desertion, and conduct prejudicial to good order, which covered a multitude of sins from insubordination to swearing. Guilt and punishment were determined by a court-martial. Punishments at Fort Larned included confinement in the guardhouse for a specified period, forfeiture of pay for a given time, reduction in rank, hard labor, confinement in a sweatbox, walking a prescribed pattern while carrying a heavy weight for several hours each day, a sentence in a penitentiary, dishonorable discharge, and branding. At one time six privates of the Tenth Cavalry who had deserted and were caught were branded on their hips with the letter "D" and sent to a penitentiary for the remainder of their enlistment periods.

Desertion was a common way to escape permanently from military duty. The number who deserted was considerable during the era Fort Larned was active, but the rate was not as high as at more isolated posts and from troops on campaigns. Some deserters were caught and punished, but most made good their escape. In 1867 in the entire army there were 13,608 desertions from a total enlistment of 53,962. In that year the Seventh Cavalry, a portion of which was at Fort Larned in the spring, lost 457 of its 865 enlisted men. Cavalry regiments experienced more desertions than the infantry because cavalrymen could use their horses to get away.

The desertion rate remained high throughout the era of the Indian wars. Secretary of War Stephen B. Elkins reported in 1891 that one-third of the men recruited between 1867 and the year of his report deserted before completing their obligations. Among the most important reasons given were general dissatisfaction with military life, the length of enlistment (all soldiers were volunteers and were in for five years), harsh troop discipline, inadequate pay and the hope of making more money elsewhere (civilian laborers at Fort Larned often made more than twice as much per month as soldiers), boresome garrison life, temptation to go to the gold mines, inferior quarters, inferior and insufficient food, and the use of troops for nonmilitary duties, such as building shelters and roads. Fear of disease during an epidemic and fear of death while on an Indian campaign also caused individuals to decide to depart and not return. Many soldiers fled while under the influence of alcohol, which was another contributing factor. That only 30 percent of those who deserted were apprehended—and many of them were returned to duty with only slight punishment—must have encouraged would-be deserters. In most instances a combination of reasons provoked a soldier to desert. Peer pressure also was involved; most soldiers deserted in groups rather than by themselves. One of the largest instances was in March 1871; forty-one men deserted following the paymaster's visit when they learned of reductions in their allowances and pay.

Desertion was not considered a serious crime in those days, but the army tried to stop it because it represented an economic loss. Deserters almost always took government property with them, some of which they might sell and some that they used for their own welfare. Also, since deserters had to be replaced it added greatly to the cost of recruiting and training new soldiers. Even so, effective means for preventing widespread desertion were not devised until the twentieth century.

Those soldiers who stayed and served out their five-year enlistment found their daily lives at the fort organized into a rigid routine. The schedule varied from season to season, but this was a typical day's routine: The first call for reveille was sounded at 6:15 A.M. Reveille was at 6:25, with assembly for roll call at 6:30. Mess call for breakfast was at 6:45. At 7:30 came stable call, when the men of the cavalry went to the stables, cleaned them, and took the horses to water. During the morning mounted drill—which usually took place two or three times a week—the horses were taken back to the stables; otherwise the animals were sent out under a mounted guard to graze at some selected place on the military reservation and within bugle call from the post headquarters.

Fatigue call was sounded at 8:00 A.M., and the men detailed from each company for such jobs as cleaning up the post, working on a construction site, loading and unloading supplies, building a road or bridge, cutting ice, and numerous other duties, were assembled on the parade ground. From there they moved on to their respective duties under the charge of a noncommissioned officer. Sick call also was sounded at 8:00, and all men who were ailing in any of the companies were sent to the

Colonel Uri B. Pearsall, Forty-eighth Wisconsin Infantry, commanded Fort Larned from October 1 to December 6, 1865. It was during his administration that a newspaper, The Plains, was published at the post. Pearsall contributed twenty dollars for his subscription, the highest pledged by anyone.

hospital to be examined by the post surgeon and treated accordingly. Able-bodied soldiers were sent to join the detail to which they were assigned. Those with minor afflictions were sent to quarters with proper medication and instructions. Serious ailments resulted in admission to the hospital for observation and treatment.

At 9:00 A.M. the new guard for the day was called to assemble at the blockhouse where he stood inspection, including performance of all or part of the manual of arms. The orders for the day and the passwords were given, and the new guard relieved the old guard for a twenty-four-hour shift. Enlisted men were assigned to guard duty by a system of rotation. They had to remain fully dressed throughout the twenty-four-hour period, during which they performed various duties. Sentinels were stationed at key posts on the fort where they were relieved every two hours. They were required to challenge intruders, call out their post number and the time every hour, and repeat the orders of the day upon the request of an officer. Sometimes guards were in charge of prisoners from the guardhouse who were required to work at some job around the post. When a guard had rest time, he spent it in the guard room of the blockhouse.

If a morning drill occurred, the fatigue recall sounded at 10:15 A.M. and the drill was called at 10:30. Drill recall was usually at noon, and dinner mess call was at 12:30 P.M. Afternoon fatigue call sounded at 1:00; if an afternoon drill took place, fatigue details were recalled and sent to drill at 2:30. Fatigue recall on days when there was no drill or from drill came at 4:30. At 4:45 came stable call again, when cavalrymen groomed the horses. When no cavalry troopers were stationed at Fort Larned there were no stable calls.

The troops were called to assembly at 5:30 P.M., with the evening dress parade and roll call at 5:40. This ended the working day at the fort. The evening mess was at 6:30. Tattoo was sounded at 9:00; this was the last roll call of the day, with the soldiers in company formation in front of their quarters. Lights were to be out when taps sounded at 9:30. Throughout the day, whenever a soldier was not answering a call, on detail, or at drill, he was free to go to his barracks or anywhere on the post not declared off-limits, as long as he was within bugle call. The Sunday routine was different, for no work details or drills were required— except for those on guard duty. There was a weekly inspection of the troops at dress parade at 9:00 A.M. on Sunday, after which the men were free until tattoo.

Men who followed the prescribed routine, reported for duty, and performed their assignments—the bulk of the troops—had free time for diversion and entertainment. Those who failed to perform were placed in confinement. How the former spent their free time varied widely: some engaged in conversation with other men in the same company; some read newspapers and books from the post library; some wrote letters and kept diaries. Others competed in contests including horse races, foot races, pitching horseshoes, and baseball games, and some of the men were fond of betting on the outcome. All types of gambling—including cards and dice—occupied many troopers. Card games also were played for pleasure. Some of the men had musical instruments and some men sang, entertaining themselves and those who listened. Practical jokes always were popular, especially upon new recruits. Sometimes a group of men would venture into drama and produce a play for the garrison's entertainment. The soldiers also were fond of dancing, although they usually had to provide their own music and dance with other soldiers. A visit to the sutler's store, and beer and whiskey drinking were other forms of diversion.

In 1865 a group of officers of the Forty-eighth Wisconsin Infantry, a regiment of Civil War volunteers anxiously awaiting their discharge from

THE PLAINS.

"WESTWARD THE STAR OF EMPIRE TAKES ITS WAY."

VOL. 1   **FORT LARNED, SATURDAY, NOV. 25, 1865.**   NO.

THE PLAINS,

Published every Saturday at Ft. Larned, Ks.,

BY THE OFFICERS AND SOLDIERS

OF THE UNITED STATES SERVICE,
STATIONED ON THE FRONTIER.

SELECT POETRY.

The Wives.

God bless the Wives,
They fill our hives
With little bees and honey ;
They ease life's shocks,
They mend our socks,
But—don't they spend the money?

When we are sick,
They heal us quick--
That is, if they should love us :
If not, we die,
And yet they cry,
And place tombstones above us.

Of roguish girls,
With sunny curls,
We may in fancy dream :
But wives—true wives—
Throughout our lives,
Are everything they seem.

"Oh for a Home Beside the Hills."

"Oh for a home beside the hills—
Where gladly leap the bounding rills—
Where sunlight dwells 'mid' fairy flowers
Which bloom, and bud 'mid green-wood
bowers :
There I would look on green vales wide,
'Mid which the gay wild waters hide.
Oh for a home beside the hills,
Where ever glide the laughing rills
A home that's bright with birds and
flowers :
"Tis there I'd live life's happy hours.

OUGHT MARRIED PEOPLE TO SLEEP TO-
GETHER?—Hall's Journal of Health,
which claims to be the highest authority in
medical science, has taken a stand against
married people sleeping together, but
thinks they had better sleep in adjoining
rooms. It says that Kings and queens do
not sleep together, and why should other
people? Think of separating a newly
married pair on a cold winter's night, because Hall's Journal of Health says so.
You can go to grass, Mr. Hall.

Military.

THE FORTY-NINTH WISCONSIN INFANT-
RY.—We learn that companies B, C and
D, of the 49th regiment, have been discharged and paid. The Colonel of the
regiment, Col. SAMUEL FALLOWS, has had
the rank of Brigadier General, by brevet,
conferred upon him, for gallant and effi-
cient services. Who's next?

Personal.

Capt. M. V. B. HUTCHINSON, Co. E,
48th Wisconsin Infantry, Post command-
ant at Fort Zarah, and Lieut. WINCHELL,
A.A.Q. M., at that Post, came up to this
post on Monday last. They report the
boys all well and everything lovely.

2d Second Lieut. CHAS. A. JOHNSON,
Co. I, 48th Wisconsin Infantry, has been
ordered to appear before the Military Commission at Washington, within fifteen days
from Nov. 6th, to answer to the charge of
"absence without leave," or stand dis-
missed the service. He has been commis-
sioned and mustered as 2nd Lieutenant of
Co. I, 48th Wisconsin, but has never
joined the regiment.

Promotions in the 48th Wis. Infantry.

Capt. Peter Trudell has been promoted
from 1st Lieutenant of Co. H, vice O. F.
Waller resigned. Date of commission
Oct. 28th, 1865.

First Lieutenant J. S. Driggs from 2nd
Lieutenant, Co. H, vice Peter Trudell
promoted.

Second Lieutenant Chas. Fowler from
1st Serg't vice J. S. Dreiggs promoted 1st
Lieutenant.

Luman D. Olin, to Captain of Co. C,
vice E. A. Bottom resigned.

Second Lieutenant John S. Kendall 1st
Lieutenant Co. C, vice L. D. Olin promoted.

First Sarg't Theophilas Dames to be 2d
Lieutent Co. C, vice John S. Herrick to
be 1st Lieutenant Co. K, vice Carver resigned.

First Sergeant Peter Mullinger, Co. K,
to be 2d Lieutenant vice Herrick promoted.

First Sargent Chas. E. Pratt to be 2nd
Lieutenant, vice Christian Amman resigned.

How we Started.

The purchase and procurement of ou
little paper was the result of a social con
vention on the evening of 18th of Octobe
when a number of us were enjoying "ou
smoke," soon after supper at the store o
our worthy sutler. A subscription pape
was immediately started, which, up to th
present time has fully realized our mos
sanguine expectations.

The following is our subscription list

Col. U. B. Pearsall,	$20.0
Capt. Chas. W. Felker,	10.0
First Lieut. S. J. Conklin, R.Q.M.	10.0
C. P. Dodds,	10.0
John F. Dodds,	10.0
Henry Bradley, Interpreter,	10.0
Frank O. Crane,	10.0
Geo. W. Crane,	10.0
Jesse H. Crane,	10.0
Capt. J. F. Hazleton,	5.0
Major J. D. Butts,	10.0
Lieut. J. G. Ball,	10.0
Lieut. J. S. Driggs,	5.0
Capt. R. Baker,	10.0
Chas. H. McKeever,	10.0
James Brice,	10.0
Surgeon L. G. Armstrong,	10.0
1st Lieut. A. V. Amet, Post Adj't,	10.0
Capt. Cyrus Hutchinson,	10.0
1st Lieut. Peter Trudell,	10.0
1st Lieut. A. B. Cady, Adjutant,	10.0
W. A. Cook, Jr.,	10.0
Capt. B. Lawrence, A. Q. M.,	10.0
Bv't. Maj. W. P. Martin, C. S.,	10.0
T. R. Curtis, Interpreter,	10.0
Lieut. M. J. Briggs,	10.0
Capt. M. V. B. Hutchinson,	10.0
Lieut Henry Felker,	10.0
1st Lieut. Don A. Winchell,	10.0
1st Lieut. W. W. Black,	10.0
Total,	$300.0
The cost of press, type, &c., at St. Louis, Mo.,	$239.0
Express charges,	99.
Total,	$338.

By the above it will be seen THE PLAI
is almost a solvent institution, and
many Western enterprises, is founded
real capital. Our thanks are due to Cha
H. McKeever, Sutler, of the 48th Wi
Inf'ty, for lending his efforts to secu
such a beautiful little Press. Also,
Messrs. M. S. Mepham & Bro., of 81,
street, St. Louis, through whom the S
Louis Type Foundry, located at No.
Pine street, were engaged to furnish us
model Press. The promptness with whi
it was dispatched reflects credit on th
enterprise of that company.

The front page of The Plains—first and last issue—printed at Fort Larned, November 25, 1865. The newspaper had three pages of news, commentary, and ads, with a blank fourth page to be used by soldiers to write their own messages to the folks at home.

82

the service, sought relief from isolation and boredom by establishing a newspaper. This first newspaper to be printed in western Kansas, *The Plains*, was conceived on October 18, 1865, by a group of officers visiting at the sutler's store. Some of them must have been involved in the printing business prior to the war. Before long they had collected three hundred dollars in advance subscriptions. A job press was ordered from St. Louis, at a cost of $239.55. The freight charges were ninety-nine dollars, making a total investment of $338.55. The printers were proud to be in debt less than forty dollars when they opened for business. The first, and as it turned out the last, edition of *The Plains* was published on November 25, 1865, just a little more than one month after the idea was raised. The Forty-eighth Wisconsin Infantry apparently took "their" printing press with them when they departed from Fort Larned in early December 1865. Thus the promise in the first issue to publish every Saturday went unfulfilled.

The paper featured poetry, commentary, military news, local news, want ads, and an editorial on the purpose of publishing the paper that stated, in part: "We are running a paper for our own amusement—for the fun of the thing.—That's all—and why not, pray tell? Why not run a paper for fun, as well as play cards or billiards, or go to a saloon or a horse race, or to hear BEECHER preach, all for fun?" The paper was both serious and "fun," a fine balance of information and wit.

In addition to an ad offering job printing of all types, including "everything in the line of letter press printing from a primer to a bible," the paper included this whimsical solicitation: "WANTED—At this office, a half dozen young ladies to learn the printing business. The foremen of this office will render all the assistance possible. None but good looking ones need apply." There was a front-page commentary strongly objecting to a recommendation in Hall's *Journal of Health* that married people ought not sleep together. The staff must have taken much pleasure in reporting on a resolution that had been offered at the Ohio Sunday School Convention: "Resolved, That a committee of ladies and gentlemen be appointed to raise children for the Sabbath School."

Among the items of local news was a report that the commanding officer and quartermaster from Fort Zarah had come to Fort Larned the previous Monday and declared "the boys all well and everything lovely" at their outpost. The day before that on Sunday, November 18, Plains Apache Chief Poor Bear and three of his tribe had come to the post under a truce flag. Poor Bear said their camp was one hundred miles south of the fort when they left it, and they came in to test the faith of the treaty signed that fall (Little Arkansas treaties). Colonel Uri B. Pearsall, commanding the post, promised to deliver food, clothing, and presents to their camp, and the Apaches went back to their people on Monday. Colonel Christopher "Kit" Carson, famous frontiersman and military officer in the New Mexico volunteers, had stopped at Fort Larned on November 12, 1865, on his way home from the treaty councils where he was one of the peace commissioners. Carson expressed his hope that lasting benefits would derive from the treaties.

The paper's weather report was concise: "The weather continues mild and unusually pleasant for November." There also was news of several prairie fires in the area during the previous two weeks. The publishers offered to accept produce of all kinds in exchange for subscriptions, and some free ad space was offered to anyone who would bring a "good-fat turkey" for Christmas dinner. The cost of ads was given as twenty-five cents per line.

Other items in *The Plains* included the arrival of a wagon-train load of Irish potatoes; an outbreak of typhoid fever that had claimed one life, that of Private James Schofield, and held another in the balance; an abundance of buffalo and wolves in the region and the fact that some soldiers were carrying on a profitable business in wolf pelts; the observation that large wagon trains were passing daily on the way to Fort Union and New Mexico with government supplies and commercial wares; and an announcement that the sutler's store had fresh pork for $.30 a pound, apples, potatoes, and onions for $7.50 per bushel, cheese for $.60 a pound, butter at $1.00 per pound, and corn meal for $.08 a pound.

The sentiment of many of the soldiers probably was summed up by one of the writers who declared: "There is no place like home—the difficulty is to get there." For the soldiers of the Forty-eighth Wisconsin Infantry, they did not have long to wait. After being delayed by a blizzard they marched out of Fort Larned on December 7 and were at Fort Leavenworth by Christmas. There they were mustered out of the service a few days later.

Life is more than work and play, even at a frontier military post, and illnesses and injuries were a constant threat to the army. The post surgeon and his hospital were important to the health and well-being of the garrison. Many health problems resulted from the environment. Crowded and poorly ventilated quarters fostered respiratory illnesses, unsanitary water induced diarrhea and fevers, mosquitoes along Pawnee Fork carried malarial afflictions, inadequate bathing facilities contributed to numerous boils, and the local prostitutes spread venereal diseases. A ready supply of liquor at the hog ranches and sutler's saloon contributed to alcoholism as well as many fights while under the influence. In 1864 it was reported that some of the soldiers were selling whiskey to Indians and demoralizing Indian women: "Dissipation, licentiousness and venereal disease prevail in and around the Fort to an astonishing extent." Some soldiers were hospitalized for gunshot wounds, although few of these were inflicted by Indians. Some resulted from fights with other soldiers and civilians, quite a few from accidents, and a few were self-inflicted. Cavalrymen often were injured by their horses. A few outbreaks of the dreaded cholera occurred, most notably in 1867 during the peak of Indian troubles after Hancock destroyed the village on Pawnee Fork.

The 1867 cholera epidemic was widespread, covering most of western Kansas, eastern Colorado, and other points along the overland routes of travel. Civilians and soldiers alike suffered from this disease, and both were treated at the post hospital. The first case at Fort Larned appeared on July 6; by July 15 a total of eleven cases were reported at the post, six of which were fatal. Only one of the dead was a soldier. After that date a quarantine hospital was set up in tents two miles from the fort, and no more cases occurred at the post. Fort Larned was one of the lucky garrisons in that epidemic, which claimed hundreds of victims in the region. The disease probably contributed to the large number of desertions on the Plains in 1867.

Cholera was a dreaded disease, but other medical problems, such as sprains, blisters, cuts, bruises, colds, influenza, broken bones, respiratory ailments, venereal diseases, and many others required the attention of the medical staff. Some of the cases were unique or unusual.

During the evening of August 5, 1868, a rabid wolf raged through Fort Larned, snapping indiscriminately at people, dogs, and such things as tents, bed clothing, and curtains. The wolf rushed into the hospital and bit Corporal Mike McGuillicuddy, Third Infantry, who was convalescing in bed. He suffered lacerations on the right arm and left hand, nearly losing his little finger. From the hospital the wolf ran to the front porch of Indian Agent Wynkoop and attacked a group sitting there. First Lieutenant John P. Thompson, Third Infantry, was bitten on both legs. The crazed animal then raced to the stables where it bit Private Thomas Mason, Tenth Cavalry, on the right foot. The wolf also ran through several other buildings and between the legs of a post sentry as he shot over the wolf's back. The guard at the haystacks finally shot and killed the wolf.

Dr. Warren Webster
1860–1861

Dr. Charles Irving Wilson
1861–1862

Dr. William H. Forwood
1867–1869

Dr. Alfred A. Woodhull
1869–1870

Pictured here are eight of the twenty post surgeons known to have been stationed at Fort Larned. Some were civilians hired to assist the army because of a limit on the number of surgeons permitted in the medical department. These men were responsible for the health of the troops, administration of the hospital, and numerous other duties.

Dr. J.W. Brewer
1871–1872

Dr. Stevens G. Cowdrey
1872–1875

Dr. A.A. DeLoffre
1875–1876

Dr. William E. Whitehead
1877

Surgeon W.H. Forwood treated the three soldiers' bites with nitrate of silver. Corporal McGuillicuddy refused to have his torn finger amputated, which proved fatal. A month after the attack he showed signs of hydrophobia; he died on September 9. The other two victims recovered.

A large dog that had fought with the wolf also died from hydrophobia. In October 1869 another rabid wolf appeared at the post, but it was shot quickly by the post sentry.

In November 1870 Sergeant David Gordon, Seventh Cavalry, was in command of an escort for the paymaster from Fort Hays to Fort Larned. He was lecturing to his men on the folly of carrying loaded weapons in wheeled vehicles when he accidentally shot himself through the knee. He was rushed the twelve miles to Fort Larned's hospital where post surgeon James N. Laing treated him. For several days the wound responded well to treatment with carbolic acid, but infection set in and sapped the sergeant's strength. When a femoral artery began hemorrhaging, necessitating the application of a tourniquet, Dr. Laing called for assistance from a fellow medical officer, post surgeon W.S. Tremaine at Fort Dodge. Together they determined that amputation was necessary in spite of Gordon's weakened condition. The operation was performed with the patient anesthetized with ether, but Gordon never regained consciousness. He died fifty hours after the operation.

In April 1871 a Private Mickey showed up for guard duty after spending too much time at the sutler's bar. When he spoke out of turn the sergeant of the guard hit him on the right temple with his rifle butt. Mickey was taken to the hospital where Dr. Laing found the man inebriated but able to communicate rationally. The soldier was placed in bed to sleep, and by evening he was insensible. He died the next day. A post-mortem showed he had a fractured skull.

Occasionally the post surgeon was called on to treat Indians. They usually would take medication provided from the dispensary but were seldom willing to undergo surgery. There were exceptions. Surgeon Forwood persuaded a young Cheyenne to take chloroform while having a finger amputated, and he operated on another Cheyenne with a broken thigh bone resulting from a gunshot.

In addition to seeing soldier, citizen, or Indian patients, the post surgeon had many other duties. The surgeon and hospital attendants were responsible for diagnosis, treatment, and surgery when necessary, and their activities included such health-related duties as sanitation, diet, examining recruits, and maintaining medical records. In addition the surgeon was required to administer the hospital, supervise all other medical personnel, dispense drugs, act as coroner, keep zoological and botanical records for the region, and record daily weather conditions.

The post surgeon had to have assistance from hospital stewards and other attendants. Their duties involved nurse care for the sick and

Post Chaplain David White, above left, served at Fort Larned, January 1–July 10, 1878. His wife, Priscilla, above right, and daughters Emma and Etta, right, lived with him at the post.

wounded, preparation of meals, provision of proper diet, changing bandages, and bathing hospital patients. These attendants usually were assigned hospital duty on a rotation basis from the companies stationed at the post, thus most were inexperienced when assigned to duty and were rotated off duty about the time they gained essential experience.

Because diet is so important to health (for example, Vitamin C deficiency causes scurvy), military posts were required to plant gardens to provide fresh vegetables. Fort Larned had gardens, but these often failed. It was only after settlers came into the area and sold produce to the post and the railroad made possible the shipment of fresh vegetables that a sufficient supply was available for the needs of the garrison.

In spite of the problems noted, it appears that the health of the Fort Larned garrison generally was good and the health care provided was satisfactory to good. During the two decades the post was active more than thirty-eight hundred persons were admitted to the hospital. Of these, forty-two died and the rest recovered and returned to duty or were discharged as being unfit for duty. The busiest year for the post hospital was 1865, with more than fifteen hundred admitted and fifteen deaths. That was also the year that the post had its largest garrison. If cleanliness could have been better enforced, the number of ailments that flourished would have been reduced considerably. The fort had no bathing facilities except for the water in Pawnee Fork until after December 1869, when the post quartermaster was ordered to construct four bathtubs for the soldiers' use.

Just as soldiers often neglected their physical well-being, they had little interest in the spiritual side of life. Fort Larned seldom had a chaplain during its existence, although visiting clergymen did hold services there on occasion. It was mainly officers and their families who attended religious services. Formal religion apparently was not very significant to most enlisted men. According to the post returns, the only chaplains stationed at Fort Larned were Benjamin L. Reed, July to September 1865; Alexander McLeod, June 3, 1871, to April 6, 1872; and David White, January 1 to July 10, 1878. One other chaplain was appointed—George P. Van Wyck in October 1870—but he never came to the post.

Life at Fort Larned during the years of Indian troubles, especially the first decade, usually was filled with some activity related to military operations off the post. Escort duty, scouting expeditions, patrols along transportation lines, guard duty with a railroad survey or construction crew, or a campaign against hostile Indians were all welcome relief from the monotony of garrison life. After the Indian wars, during the last decade

of the fort's occupation, life at the post was even more routine and bor-
ing. The size of the garrison was small, few opportunities arose for field
service, and much of the work involved the maintenance of a post that
had begun to fall into disrepair. Even so, life was easier for the soldiers.
They ate well, had access to the rest of the nation via the railroad and
telegraph, and could spend leisure hours as they pleased: reading, writing to family and friends back home, fishing in Pawnee Fork, or visiting
nearby settlements. During those last years the post was much more like
a small village than a fort.

9

The End and a New Beginning

By the early 1870s it was obvious that the circumstances that had led to Fort Larned's founding had passed and the reasons for its continued existence were few. Wagon trains and stagecoaches had been replaced by railroads. Indian agencies and distribution points for Indian annuities had been moved to present-day Oklahoma along with the Indians. The troops left at Fort Larned had few tasks beyond post maintenance. General Sheridan suggested in 1872 that Fort Larned had served its purpose and recommended its abandonment because Indian troubles were no longer a threat to the area. General John Pope, department commander, concurred in that decision, but Kansas Governor James Harvey appealed directly to the War Department to keep the post active. Harvey declared that railroad crews and new settlements in the region needed the assurance of protection from Indians who still broke away from their reservations to raid from time to time.

Fort Larned was not abandoned in 1872, but from that time until it was deactivated the garrison slowly was reduced in size, and the military value of the post decreased each year. The troops there did provide assistance to settlers after the grasshopper invasion of 1874, when they transported supplies to destitute settlers and distributed government food and clothing from the post.

As settlements around the military reservation grew and the civilian population in the region increased, the fort garrison was reduced to little more than a guard for government property. Portions of the reservation

This 1886 view of old Fort Larned shows the early conversion of the post to a ranching operation by the Pawnee Valley Stock Breeders' Association. The blockhouse still was standing, although this is the last known photo in which it appears. Some of the storehouses have been converted into barns. At least two windmills have been added to pump water for livestock.

were leased to ranchers, an indication that the land was useful and soon would be desired for purposes other than military. Few objections were heard in 1878 when General Pope again recommended the abandonment of Fort Larned. On July 13, 1878, the garrison was sent to Fort Hays. Many of the military supplies were transferred to Fort Dodge. A small detachment was left to guard the abandoned buildings and remaining government property.

Settler interest in the 10,240-acre military reservation increased after the troop evacuation. Kansas Senator Preston B. Plumb introduced legislation that became law in 1882 providing for the transfer of the reservation to the General Land Office, survey and appraisal of the land, sale of the land in quarter sections (160 acres) under the provisions of the preemption law, and the sale of the section (640 acres) containing the post buildings in one unit either at auction or by private sale as the commissioner of the General Land Office deemed to be in the best interest of the nation.

A portion of the reservation became part of the federal land grant to the Atchison, Topeka and Santa Fe Railroad; a portion was sold directly to settlers by the General Land Office; a portion was sold through H.M.

Bickel and Henry Booth, land receivers with offices in the town of Larned; and the section with the buildings was sold at public auction. The auction, which took place in Larned on March 13, 1884, resulted in a high bid of $11,056 ($ 17.271/2 per acre) by representatives of the Pawnee Valley Stock Breeders' Association. After an initial default in payment, the purchase price as bid eventually was paid, and the land was patented to the association on June 12, 1885.

The association utilized the old fort as part of its ranching operation, but financial difficulties resulted in the loss of the land and buildings. The former fort and other land was mortgaged to an investment company in Kansas City, Missouri, for ten thousand dollars. The mortgage was sold to George D. Wilbur of Boston. When the Pawnee Valley Stock Breeders' Association defaulted on May 1, 1891, George Wilbur sold the land to Charles Wilbur, who had been a member of the association. On January 5, 1893, Charles Wilbur sold it to Johanna Frorer of Illinois. In 1902 Frorer sold approximately three thousand acres, including the fort buildings, to Edward Everitt Frizell. The Frizell family developed this land into a successful farming and livestock operation, with the post buildings preserved and utilized.

The commanding officer's quarters became a family residence. The other officers' quarters housed employees. The barracks later were connected together and remodeled into a large barn. The bakery and shops building and the new commissary storehouse became machine shops. The old commissary storehouse and quartermaster storehouse were converted into barns and storage facilities. The parade ground was fenced for livestock. These were fortunate developments because the result was the preservation of the stone structures, making Fort Larned one of the best preserved frontier forts in the American West.

Changes were made to the fort, some of them before the Frizell family acquired the land. In 1886 the military cemetery was abandoned; this was the second cemetery at the post, located approximately one-fourth mile northwest of the quadrangle. The sixty-eight known graves were exhumed and the remains transferred to the national cemetery at Fort Leavenworth. It is believed that a number of unknown graves were left behind. Sometime after 1886 the blockhouse was torn down. Some of the stone may have been used in the remodeling of some of the other buildings, and some may have been used for construction elsewhere. Other buildings of less permanent construction were razed. Windmills were erected to pump water. Most of the native grassland was cultivated.

E.E. Frizell's son, Edward D. Frizell, and grandson, Robert R. Frizell, continued the efforts to preserve the historic buildings. They saw the

This aerial view of the Frizell's Fort Larned Ranch illustrates how the fort was converted to meet the needs of that operation. The old oxbow of Pawnee Fork was turned into a pond to store water. The barracks were connected and hay lofts added to make one of the largest barns in the region. Corrals, silos, roads, trees, and fences were added to enhance the utility of a farm and ranch. Note, too, that most of the land was cultivated and planted to crops such as alfalfa, milo, and wheat. Highway 156 borders the ranch in the background. All of the farm additions and alterations, with few exceptions, have been removed since the National Park Service restored the fort to its original state.

value of their ranch as a historic site and possible tourist attraction. Signs were placed along U.S. Highway 156 to welcome visitors to the old fort. In 1955 a representative of the National Park Service, Merrill J. Mattes, investigated the site and recommended it for designation as a national historic site because of the good condition of the original buildings, the site's relevance in commemorating the Santa Fe Trail's famous history, and its proximity to a nearby federal highway, which ensured easy access to the site. Congressional interest was spurred by this report.

Further attention was drawn to the post in 1957 when the Fort Larned Historical Society was founded under the leadership of Larned newspaper editor Ralph Wallace. The Society's purpose was to develop a tourist attraction at the fort and seek federal acquisition of the site for historical preservation. The old fort opened to the public in 1957, and a fine museum collection and visitor's center were developed. A special celebration of the fort's centennial was held in 1959. In 1961 the Department of Interior designated Fort Larned a national historic landmark. In 1964 Congress authorized the National Park Service to incorporate Fort

When the Fort Larned Historical Society opened the site to visitors, this was the appearance of the west end of the west enlisted men's barracks; it had been remodeled into a barn for the ranch and served, at the time of this photo in 1965, as the museum and visitor's center. This building, now restored to its original appearance, currently houses offices, the visitor's center, and fine museum exhibits.

Larned into the park system as a national historic site. The land and buildings were purchased by the government in 1966. The Fort Larned Historical Society operated a museum and most interpretive services at the site until 1973. In 1974 that organization opened the Santa Fe Trail Center three miles east of the fort, where a museum and research library dealing with all phases of Santa Fe Trail history are open to the public.

The National Park Service is restoring the extant buildings as nearly as possible to their original condition. When completed these structures will appear essentially as they did in 1868, with the exception of the east barracks, which will include the hospital as it did after 1871. This restored fort, with its authentic museum exhibits and living history demonstrations, commemorates the history of a frontier military post on the Santa Fe Trail that assisted travelers and traders, stagelines and railroads, miners and cowboys, buffalo hunters and farmers, merchants and other settlers who expanded the nation's boundaries and built the West. Here visitors from home and abroad can better understand and appreciate this important heritage.

A 1993 aerial view of Fort Larned after restoration.

These historical exhibits in the museum and furnished buildings at Fort Larned provide visitors with a feeling for the era and an understanding of the events experienced by the soldiers stationed there from 1859 to 1878.

National Park Service personnel and volunteers offer living-history demonstrations, such as these of black cavalrymen, above, and a platoon detail, below, to help visitors appreciate and understand life at this frontier army post.

Appendix

Officers frequently changed, and some served several different times. Commanding officers seldom served more than a few months at one time. The final two were the only ones to serve longer than one year at a time.

Officer and Regiment	Year(s)
Captain George H. Steuart, First Cavalry	1859
Lieutenant David Bell, First Cavalry	1859–1860
Captain Henry W. Wessells, Second Infantry	1860
Lieutenant Lloyd Beall, Second Infantry	1860–1861
Captain Julius Hayden, Second Infantry	1861–1862
Captain Daniel S. Whittenhall, Second Kansas Cavalry	1862
Major Julius G. Fisk, Second Kansas Cavalry	1862
Captain H.N.J. Reed, Ninth Kansas Cavalry	1862, 1863
Lieutenant Colonel Charles S. Clark, Ninth Kansas Cavalry	1862
Captain Jacob Downing, First Colorado Cavalry	1862
Lieutenant William West, Second Infantry	1862–1863
Lieutenant Watson D. Crocker, Third Wisconsin Artillery	1863, 1864, 1865
Captain James W. Parmetar, Twelfth Kansas Infantry	1863, 1864
Colonel Jesse H . Leavenworth, Second Colorado Infantry	1863
Captain W.H. Backus, First Colorado Cavalry	1864
Major Scott J. Anthony, First Colorado Cavalry	1864
Captain E.A. Jacobs, First Colorado Cavalry	1864
Captain Thomas Moses, Jr., Second Colorado Cavalry	1865
Captain Theodore Conkey, Third Wisconsin Cavalry	1865

Captain C.O. Smith, Fifteenth Kansas Cavalry   1865

Captain Thomas J. Malony, Jr., Second Cavalry   1865

Colonel Uri B. Pearsall, Forty-eighth Wisconsin Infantry   1865

Major Hiram Dryer, Thirteenth Infantry   1865–1866

Lieutenant James Cahill, Second Cavalry   1866

Major Cuvier Grover, Third Infantry   1866

Captain Henry Asbury, Third Infantry   1866–1867, 1868

Major Meredith Helm Kidd, Tenth Cavalry   1867–1868

Captain Nicholas Nolan, Tenth Cavalry   1868

Captain Daingerfield Parker, Third Infantry   1868, 1869, 1870

Major John E. Yard, Tenth Cavalry   1868–1869

Captain James Aiken Snyder, Third Infantry   1869–1870, 1871

Lieutenant Charles Louis Umbstaetter, Third Infantry   1869

Captain Verling Kersey Hart, Third Infantry   1870–1871

Major Richard Irving Dodge, Third Infantry   1871

Captain George Edward Head, Third Infantry   1871

Major James P. Ray, Sixth Infantry   1871–1872

Captain Henry B. Bristol, Fifth Infantry   1872–1873

Captain Simon Snyder, Fifth Infantry   1872, 1873–1874

Captain William John Lyster, Nineteenth Infantry   1874–1877

Captain Jacob Hurd Smith, Nineteenth Infantry   1877–1878

Further Reading

Berthrong, Donald. *The Southern Cheyennes* (Norman: University of Oklahoma Press, 1963).

Brown, Dee. *The Galvanized Yankees* (Urbana: University of Illinois Press, 1963).

Custer, George A. *My Life on the Plains or, Personal Experience with Indians* (Norman: University of Oklahoma Press, 1962).

Duffus, R.L. *The Santa Fe Trail* (New York: Longmans, Green and Co. 1930).

Jones, Douglas C. *The Treaty of Medicine Lodge* (Norman: University of Oklahoma Press, 1966).

Leckie, William H. *The Buffalo Soldiers: A Narrative of the Negro Cavalry in the West* (Norman: University of Oklahoma Press, 1967).

Leckie, William H. *The Military Conquest of the Southern Plains* (Norman: University of Oklahoma Press, 1967).

Moorhead, Max L. *New Mexico's Royal Road: Trade and Travel on the Chihuahua Trail* (Norman: University of Oklahoma Press, 1958).

Oliva, Leo E. *Soldiers on the Santa Fe Trail* (Norman: University of Oklahoma Press, 1967).

Rickey, Don. *Forty Miles a Day on Beans and Hay: The Enlisted Soldier Fighting the Indian Wars* (Norman: University of Oklahoma Press, 1963).

Stallard, Patricia Y. *Glittering Misery: The Dependents of the Indian Fighting Army* (Fort Collins: Colo.: Old Army Press, 1978).

Utley, Robert M. *Frontier Regulars: The United States Army and the Indian, 1866–1891* (New York: Macmillan Publishing Co., 1973).

ACKNOWLEDGMENTS

The author wishes to thank the following friends for help, information, and encouragement in the preparation of this material on Fort Larned: Timothy A. Zwink, George Elmore, B. William Henry, William E. Unrau, Everett M. Brown, Joseph W. Snell, and the excellent staff at the Kansas State Historical Society. Most of the research was conducted in records contained in the National Archives, and I am grateful for the microfilm that institution provided. A special note of thanks is due my wife, Bonita, without whose encouragement and help this would never have been published.

ILLUSTRATION CREDITS

Frontispiece: Bonita M. Oliva; facing title page: Department of the Army, U.S. Military History Institute; 2, 4 KSHS; 8 Colorado State Historical Society; 9, 10 KSHS; 12, 13, 14–15 Fort Larned National Historic Site; 18 Colorado State Historical Society; 21, 22, 23 KSHS; 24 original blueprint property of the National Archives; 26 KSHS; 27 Colorado State Historical Society; 28, 29, 30–31, 32, 33 KSHS; 34 Fort Larned National Historic Site; 35, 36, 37 KSHS; 38 (top) Fort Larned National Historic Site; 38 (bottom), 42 KSHS; 44 (top left) Western History Collection, University of Oklahoma; 44 (top right, lower left, lower right) KSHS; 45 (left) Smithsonian Institution; 45 (right), 46, 47, 48, 49, 50, 51, 52 KSHS; 56 (top left) U.S. Military Academy Archives; 56 (top center, top right, bottom left) Colorado State Historical Society; 56 (bottom right) Fort Larned National Historic Site; 57 (top left) Fort Larned National Historic Site; 57 (top right) U.S. Signal Corps, National Archives; 57 (bottom left, bottom right) Fort Larned National Historic Site; 58–59 KSHS; 59 (right) Fort Larned National Historic Site; 60 (top left) Harold G. Jasperson; 60 (top center) Fort Larned National Historic Site; 60 (top right) Kansas Collection, University of Kansas Libraries; 60 (bottom left) Fort Larned National Historic Site; 60 (bottom right) Ellis County Historical Society; 61 KSHS; 62 (top left) Library of Congress; 62 (top center, bottom left) Fort Larned National Historic Site; 62 (top right, bottom right) Colorado State Historical Society; 62 (bottom center) U.S. Signal Corps, National Archives; 63 (top left) U.S. Military Academy Archives; 63 (top center, bottom center, bottom left) Fort Larned National Historic Site; 63 (top right, bottom right) Colorado State Historical Society; 64 (top left, top right) Fort Larned National Historic Site; 64 (top center) U.S. Military Academy Archives; 64 (bottom), 68, 69, 70, 72 (top, bottom right) KSHS; 72 (bottom left) Fort Larned National Historic Site; 73, 74, 80, 82 KSHS; 86, 87, 89 Fort Larned National Historic Site; 94, 96 KSHS; 97, 98, 99 Fort Larned National Historic Site.

This publication has been financed in part with federal funds from the National Park Service, a division of the United States Department of the Interior, and administered by the Kansas State Historical Society. The contents and opinions, however, do not necessarily reflect the views or policies of the United States Department of the Interior or the Kansas State Historical Society.

This program receives federal financial assistance. Under Title VI of the Civil Rights Act of 1964, Section 504 of the Rehabilitation Act of 1973, and the Age Discrimination Act of 1975, as amended, the United States Department of the Interior prohibits discrimination on the basis of race, color, national origin, disability, or age in its federally assisted programs. If you believe you have been discriminated against in any program, activity, or facility as described above, or if you desire further information, please write to: Office of Equal Opportunity, National Park Service, P.O. Box 37127, Washington, D.C. 20013–7127.